SASQUATCH

THE PROOF

BY

THOM CANTRALL

Table of Contents

I.

It Happened One Night

It was warm in Georgia during the day even this late in the year. When evening fell, however, the Fall air cooled and a fire in the central pit was a welcome thing.

It was mid-October of 2014 when more than twenty of our gang gathered at a county park in the western part of the Peachtree State for a week in camp. We had come from near and far to meet here this stellar week… some from just up the road and others from about as far away as possible and still be in this country and keep our feet dry! We even had visitors from far, distant lands with us on this wonderful week.

This day had been eventful, with visits from knowledgeable

people who lived in the area. An invitation had been proffered, and accepted, to visit their lands and see for ourselves what was happening there. They told of the many incidents they had been having with the sasquatch people and at least two of our people had been privileged to have their first encounter with the large hominid that day on that excursion.

As the sun dipped to the west, Jonathan plied his trade and kindled a nice fire for us we began gathering around it with high spirits and glad hearts. Our arrangement around that fire was roughly as illustrated below. At this remote date, I cannot be totally sure of the exact placement of everyone around it, but this is very close. Also, there may be one or two discrepancies in those in attendance, but not more than two. There were eighteen people present, but I may have

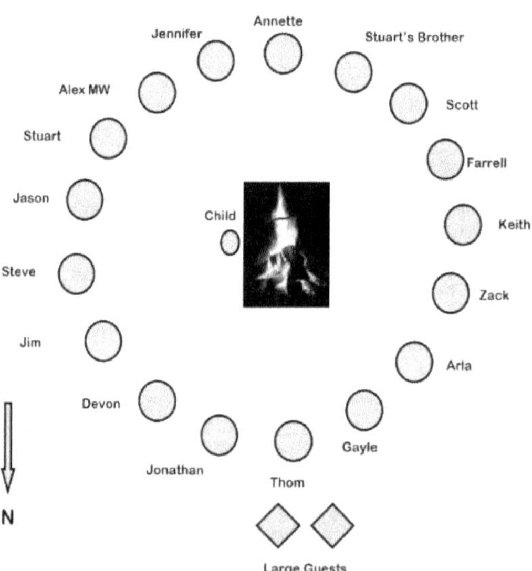

the names wrong on one or two of them on east side of the fire.

The mood was light with an animated conversation going that recounted the events of this great day when I, looking directly south across the fire saw Annette suddenly look stricken. My immediate thought was, "Oh, my gosh, there is something terribly wrong with her… she is ill!"

Her face was suddenly very pale. Her eyes were open inordinately far and her lower lip was quivering like she was

about to cry. As I watched her, her right arm began to rise and she eventually pointed directly above my head with a fixed glare to her eyes. From her quivering lips came a stuttering… "Look THERE…"

Quickly, I looked back over my head to see not one, but TWO sasquatch people, one much larger than the other, suggesting a youngster accompanied by an older individual, standing directly behind me and observing our campfire from the positions indicated on the layout above. The

outburst of activity from around the fire with someone calling out, "Guys, guys, we have company…" caused the pair to immediately turn and fade into the forest from which they had emerged prior.

In the short time I had to view them from an inverted viewpoint, I judged the larger one to be approximately seven feet tall and the smaller one to be in the five-foot range.

Immediately, Arla rose, followed by, first, Jonathan, then Keith and young Zach. The four of them followed after the big guys for a short way when they came upon a sasquatch in the brush, whereupon they all stopped, Arla and Zach in the lead, Jonathan directly behind her and Keith in the rear. Immediately, Arla began talking to the fellow, whereupon, he stood up to his full height and was joined by the two who had been with us at the fire.

After holding this tableau for a short time, the three sasquatch people slowly moved on into the heavier brush and disappeared. The four people returned to the fire to report what had transpired. It was young Zach who said it best at the time… "I've seen more bigfoot on

one weekend that 'Finding Bigfoot' has seen in three years!" …You know, he's right!

II.

Can We Trust Our Eyes?

About forty years ago, Dr. Steven Covey, author of the Best Seller, "Seven Habits of Highly Effective People", penned a book that was of great significance to me. This volume was entitled "The Divine Center". While it's subject might not be pertinent to everyone in our world, there is one treatise within it that is extremely relevant to us all.

Dr. Covey advances the concept that we see things differently, depending on what we have perceived prior in our lives. To illustrate this point, he used a concept wherein he introduced us to a picture, then had another view a different picture. This done…

Well, I think it'd be easier to show you using Dr. Covey's pictures… This will take two people to work, and both must cooperate with the instructions… If both are ready to attempt this, Person 1 should look at the picture on page 17 for a few seconds. Person 2 does not look at the picture on page 17. This is important, only Person 1 is to see the picture on page 17. When he is complete, Person 2 should look at the picture on page 36. Again, only Person 2 is to see the picture on page 36. When Person 2 has seen it for a few seconds, both persons go to page 97 and view the picture there.

With both of you viewing the picture on page 97, each should describe to the other what he sees. Do not point to the picture nor use hands to outline the figure. Simply describe to each other what that picture shows.

As should be readily apparent, we see two different things in this picture, do we not? Why? Simply because we had different viewpoints… i.e., we were reared differently, in different environments, that rearing being the previewing of a picture that predisposed us to a particular point of view.

It is important to remember that life does that to us as well and this is one of the things that causes "eyewitness" evidence to have an unreliable face to it! How was that witness reared? What were his experiences as compared to ours? These are important things to know if we are to arrive at the truth, are they not?

Opening Argument

On Proof, Respect and Fools

"Show us the proof," the fellow name Rick A. demanded. "Sixty years of research and you don't have an ounce of proof," another said vitriolically.

Thus is life in the world of he who investigates and researches the sasquatch people.

I would like to begin by stating that these and all others of their ilk are dead wrong! If I had no proof, I would have long since given up in disgust. I have little, in the way of patience, and do not suffer long periods without positive results well. The subtle difference is, I have sufficient proof for ME... if that is not enough for these others, I am sorry, but that is not my problem, is it?

It has to be remembered that no one can "PROVE" anything to another. We can only offer evidence. It is up to each individual to determine what level of evidence provides proof in their own mind. That level will be different in each person... hence, our legal juries are made up of twelve of our peers. Why? Because of that very tenet... EACH of the twelve will require a different level of evidence to be convinced of the guilt of the individual… and it requires all twelve to agree in order to convict, does it not?

Evidence comes to us in various classes, some more desirable than others. It is essential that we understand this and that we treat each class with the respect due it.

The first and most desirable evidence is "Empirical Evidence." This form of evidence is repeatable and measurable. "I found a footprint that measured fifteen inches in length and made a cast of said footprint." This is something we can measure and test… Anyone can come forth and do their own measurements. It is quantifiable and measurable. The only thing to be suspect here is the reputation of the person presenting the cast as evidence. Is he reliable, or does he have a reputation for creating false evidence? If Ray Wallace, the infamous track hoaxer of the late 1950's in northern California presented it, I would immediately question its authenticity. If Dr. Jeff Meldrum presented it, I'd accept it without question.

The second class of evidence is "Eyewitness" evidence… What people saw. This class has to be scrutinized diligently for five different people

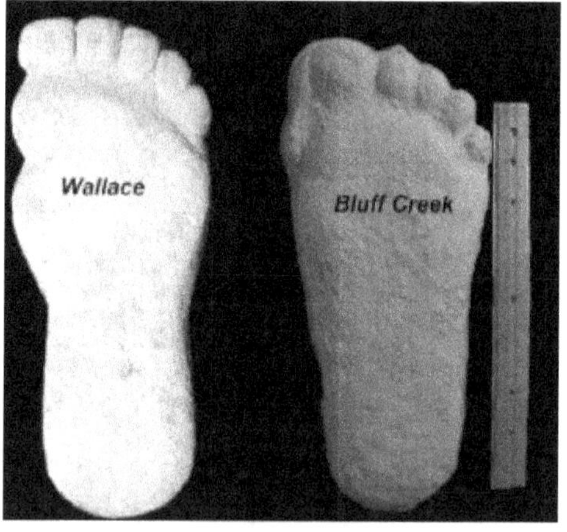

Wallace Print… Notice toes

witnessing an event may well see it from five different viewpoints and certainly from five different personal backgrounds.

The illustration from Dr. Covey on page eight, "Can We Trust Our Eyes" is exactly what we are discussing here.

Some years ago, a psychological test was done with a group of Inner-city ghetto youths wherein they were shown a teddy-bear and asked to identify it. The vast majority of the children identified it as a rat! Why? Because they knew what rats were, they lived with them. They had no experience with a teddy-bear, having never seen one. Therefore, it becomes imperative to determine the biases and teachings of the eye-witness. Were they scared by the vision? Were they calm and if so, why were they not frightened? Level of education becomes important in many cases as well, as that can determine the witness's level of comprehension of that which they are seeing.

The third class of evidence is "Anecdotal" evidence. This, basically is a story. "The mountain was really steep and, while we didn't see anything, we heard what we knew was a..." Again, we have to rely on the reputation of he who is telling the story. Is he reliable or is he prone to not letting the

truth stand in the way of a good story? This class of evidence is highly suspect because it does rely so heavily on the proclivities of the person relating the anecdote.

The fourth and least reliable form of evidence is "hearsay". In fact, it is so unreliable as to not being allowed in court cases. Basically, hearsay is just that… what one heard and is repeating. "My brother-in-law said…" is a form of this type. "My neighbor told me that she saw this large thing…" As can be seen, he main reason this is inadmissible, in most cases, is because the individual source of the statement is not available for cross examination. All we have to rely on is that portion related by the person relating the statement.

These four classes of evidence are presented here in the order of their acceptability with empirical evidence being the most reliable and hearsay evidence being the least reliable, even to the point of being unacceptable on its face.

This said, is there good evidence? We have excellent videos… we have superb photos… neither of which will ever be "proof" in and of themselves if, for no other reason than that the editing software is too good. Argument rages until all reason is lost and insanity prevails. Witness the Patterson-Gimlin Film… a minute of film that has sparked a half century of often bitter rancor. Yet, the professionals… those who know costuming… those who know film and cameras, those who know the intricacies of making video features and those who understand anatomy, physiology and anthropology are ignored while a fool brother-in-law is cited ad infinitum… ad nauseum… and my brother-in-law is a total idiot. The consensus among all these experts is, "It would be impossible to create a suit in 1967 capable of doing the things that film showed. That the material required simply did not exist at that time!" But that brother-in-law said it was a man in a suit, so that's what we will believe. Insanity? I think, perhaps.

As evidence, I have a plethora of footprints... I have structures I have vetted to my satisfaction to be authentic... I have an audio, aural language recorded, vetted by a university and expounded on by a professional crypto-linguist. I have physical evidence of a written language... a written language that is found across North America in profusion. There are casts of butt prints showing anatomical features such as labia, anus, hair follicles and even scrotum images. We have examples of defecate of immense size that does not match anything known or tested... we have hair samples that match nothing known to man... but my brother-in-law says it's a myth... so we believe him... and he's an idiot...

These are examples of hard, physical evidence... things we can see, feel, examine and test. From there we can move to the slightly more esoteric. We can venture into the realm of the subjective. Testimony...

In all cases of testimony, the underlying necessity is that the person testifying be believable. He has to have a degree of credibility to be trusted. One need only watch a lawyer in cross examination to understand the degree of importance to this tenet...   Let a testimony be from someone of questionable character and that fact alone will negate all value of any such testimony in very short order without the most special of circumstances being in order.

In conclusion, let me say that, yes, there is solid evidence… ample certainly, to get a conviction in a court of law. But "Proof" is a very personal thing with each of us requiring his own level and class of evidence being required. Even as in a court of law, there are those who will never be convinced, but unlike that court of law, we cannot dismiss that type as unsuitable to sit in judgement. Alas, our idiot brother-in-law will hold sway.

One very negative thing the internet, especially sites like Facebook or Twitter, has done is to give voice to idiots. Those who, heretofore, had to stand on a soapbox in the park to make their message heard now have a ready-made audience and they do make the most of it… Beware of fools in large numbers…

Test Photo #1

1.
1958 – The Jerry Crew Story

The sun had not yet made its appearance this morning in late August, 1958. The stars had faded from their prior glory, but still could be seen here and there to belie the coming heat of a summer day in the mountains of far northern California. Slowly, by threes and twos, the crew building the roads that would facilitate the moving of the vast storehouse of Douglas fir growing here to the sawmills of the flatland were making their way back to work. It was Monday... and Monday meant they

Jerry Crew - Crew Family Archives

were driving in from town after an all too short weekend at home with family. Most of these men stayed in one camp or another, some down on the paved road, where families often joined them, on the Klamath River, while others utilized more rustic accommodations found by a convenient wide spot here or there along the stream they knew as Bluff Creek.

It was but the matter of a few moments for those in this weekly migration to drop off the items brought from town to sustain them for the coming week. Some changes of clothes... food, of course and perhaps something liquid to make the nights more sustainable. This task complete, it was on to the job site, now some twenty-five miles in off the rough pavement of State Highway 96 on a freshly built road that climbs steeply out of the Klamath River basin in switchbacks. Past Blue Lake, beyond the Cooper Ranch, Bee Lake all the way to their equipment and materials storage dump at Louse Camp and more... on they drove until reaching the open dirt that told them they were home... This was their pumpkin patch... the road they were building. Twenty-five miles is a long way to travel on two ruts in the dust! It was not the work of fifteen minutes and a cup of Starbucks to gain this job... it was more like three times that long and it took a brave (or foolish) man to attempt to drink hot coffee on this route. The chances of wearing that coffee were much higher than ever gaining a taste of it!

Eventually, however, the time came to exit the pickups that brought them in and start up the heavy equipment that would be their home for the next ten hours or more. On this morning, however a surprise was awaiting this gang of experienced mountain people... a surprise of gigantic proportions.

Gerald Crew was a cat skinner for Ray Wallace's

construction company who had contracted for the building of the roads into and the logging of a stand of timber owned by the U.S. Forest Service as part of the Six Rivers National Forest and purchased by Block and Company, the prime contractors on this job. On this morning, Jerry, as he was known to all who knew him, walked in the soft dirt track to start his tractor and let it warm up while he did the daily maintenance routine of greasing all the zerks and checking fluid levels on his machine. Jerry was a very experienced and well-respected man and had been building roads in this country for long enough to know that the tracks he saw in the soft dirt around his machine were not normal.

The man's first reaction was one of anger. Who in the world was up here playing silly pranks on them by making all these tracks to fool them into thinking something that large had been here? Then he stopped to consider... no one had been here all weekend... there were no other tracks in the soft dirt of the freshly cut road bed. No one had even stayed in the camps back down the mountain and how would they have gotten here anyway without leaving, at least, footprints?
Jerry called the other members of the crew together and amid the hoots and haws of derision he was hearing from some of them, others made the same interpretations... these

were large tracks and there were no others... no one could even imagine how someone could make a sixteen-inch footprint and then walk down the road leaving a track-way with a measured stride of from forty-six to sixty inches? It just didn't make sense to these men that this could be happening.

As a forester, I was taught a method to develop a standard "pacing" step. I practiced this method until it became so second nature to me that I could measure

extremely long distances to very close tolerances by pacing the distance and recording the number of steps it took to traverse the distance. My "stride," heel of left foot to heel of right foot in this system was exactly thirty inches. My "pace," heel of left foot to heel of left foot, or a double step, was exactly sixty inches. I have measured distances of over a mile using this discipline and been within a few feet of my goal at the end... often too close to make adjustments. I am six feet and four inches tall and my "stride" as the term is used here and as Dr. Sanderson used it in his writings, is thirty inches. The being who left that track-way around Jerry Crew's cat had a stride of, nominally, over fifty inches.

Oddly enough, the only crew members not surprised to see this strange track-way were the Indians who were part of the crew. The land they were on was very near both the Hoopa and the Yurok Indian Reservations and many of those men worked on the construction and logging crews in their area. These men just looked at the tracks, smiled and returned to their duties.

It was James Crew, Jerry's nephew and fellow crew member who suggested they follow the tracks to see if they could learn more of his perambulations. It was determined that the subject who left those tracks had not entered the scene on the road, but had descended a very steep slope to the roadway. That mountain slope was determined to be at an angle of seventy-five degrees... a 375% slope... meaning that it rose three and three-quarters feet for every foot it moved horizontally. Believe me, that is steeper than a man would want to walk. The tracks left the slope and proceeded to Jerry's cat where it circled it before proceeding back down the newly cut roadbed. It then cut across the road to the opposite side to the one it had been walking and dropped off another bank stated by Dr. Ivan Sanderson, "To be steeper than the slope it had descended in attaining the road initially..." though how much steeper, I could not imagine.

To illustrate exactly how steep this slope was, do the following. With a piece plain paper, mark a point. From that point, measure horizontally one inch and make another point... measure vertically for three and three quarters inches and make a third point. Now, simply connect your first point with this third point and the resulting line represents the slope traversed by that creature. Now each of us should ask ourselves how we would do trying to negotiate that grade. In comparison, no Interstate Highway will have a grade exceeding eight

percent. No logging road on public land will ever exceed sixteen percent. A normal stairway in a house should never exceed ninety-two percent grade. This slope was almost four times that steep. To match this slope in a step of stairs in a house, the standard steps would be nearly a full YARD above the previous one!

It was stated that the only time the measured stride varied significantly was when an obstacle presented itself to the individual's free travel. The size of the track and the stride reported, by application of the anatomical standard for primates, would indicate an individual of a nominal eight feet and eight inches in height and in excess of seven hundred seventy-five pounds in weight. That, as one can readily see, is larger than any human living.

By this time, the machines were all warmed up and ready to run, so work became the order of the day and the tracks were abandoned, though not forgotten. Talk persisted as to what they could possibly be... what could possibly have made them and discussions of other tracks naturally came up. This was not the only, nor even the first such occurrence of this type.

Over the prior years, such things had been reported on a regular basis. Just a few days prior, a report very similar had come from a crew building roads in the Korbel area, just a few miles south of this crew.

More importantly, this was not the last such occasion. This fellow returned on a semi regular basis to inspect the work site, leaving his large tracks in the soft soil to confound and confuse the men working there. Every few days, he would return, only to disappear again and not come back for in interval.

One of the staunchest skeptics among the crew began telling his wife in camp what they were finding on the mountain... at first she just did not believe the stories he was telling her, but as he became more convinced and less skeptical, she became more confused... so much so that in late September Mrs. Jesse Bemis wrote a letter to Mr. Andrew Genzoli, a columnist on the Humboldt Times newspaper in Eureka, California, stating that there was talk all around the job about "Wildmen" living in the mountains of Humboldt and Del Norte Counties and was there any truth to this? Does a wild man really live in these mountains?

Mr. Genzoli stated that he first regarded Mrs. Bemis' letter as a prank but as it lay on his desk, it began to look more and more interesting, so he finally decided to run it in his column on September 21, 1958. The response was not at all what he expected. Although he did get a percentage of reader's responses expressing hate and vitriol for broaching the subject, the majority of the response was in support of her question and contained many stories from others who had experienced this phenomenon.

On October second and third, our big footed friend returned to the Ray Wallace job but this time, Jerry Crew was prepared and he used plaster of Paris to make plaster castings of the big guy's tracks. The big guy returned on the fourth as well, but by this time, Mr. Crew had traveled from the job site, it being the weekend, and had taken his track casts to Eureka to show a friend.

Word got to Andrew Genzoli that Jerry Crew was in town with these casts and he invited him to get together to discuss what had been happening. Mr. Crew and Mr. Genzoli met and Jerry even allowed himself to be photographed with his track casts. He did, however, refuse to smile in any of the pictures because he did not want to the photos to "portray that I might be flippant or less than serious about what he was showing here." Andrew Genzoli satisfied himself that the results of his investigation into Jerry Crew's background was, in fact, true and that Jerry was a non-drinking active member of his church who enjoyed a well-deserved reputation as a sober, reliable and truthful person.

Following this interview, Andrew Genzoli published a feature article that appeared as a headline on October fifth, 1958 in the Humboldt Times newspaper. This article was picked up by the wire services and appeared in newspapers around the world. In this article, Mr. Genzoli called the creature, "Bigfoot", the first time that name was applied to the being in print.

Dr. Sanderson, in his November 1961 report that appeared in "True" magazine stated concerning this event: "...There, (Bluff Creek, California) enormous footprints turned up night after night all over a new road being bulldozed into a wilderness area not one hundred miles from the town of Eureka. They were inspected by several hard-boiled and highly practical-minded bulldozer operators, loggers and road-engineers and even by press photographers. They were up to twenty-two inches long, appeared night after night out of impenetrable forests, went up impossible slopes, meandered around machinery left parked at night, and wandered off back into the wild with sixty-inch strides. They caused a great stir, which prompted some enquiry. This brought to light the fact that such things had been reported off and on for a century all over the area and as far away as Idaho, Oregon and Washington. Further, they linked up with similar sightings in British Columbia."

It must be noted here that Dr. Ivan T. Sanderson, in the decades of the Fifties and Sixties was a world-renowned Paleontologist and Mammologist. In that time period he had written thirteen books on Mammals that were the base line for teaching across the world. In addition to this, he was a regular contributor to such magazines as "True" and "Argosy" ... men's magazines dedicated to the dissemination of information from the scientific world to the layman. As can be imagined, his dissertations are factual and interesting.

At the time of the incidents related, the company's owner had been away and out of the country. He had been in Costa Rica, Central America assessing the possibility of helping that nation access and harvest some of the very valuable hardwoods species that were indigenous and found in great numbers in the mountainous jungles so prevalent there. Ray Wallace returned to his job site during the second week in October, 1958 to find out all that had been happening in his absence. To say that his job-site was bedlam would have been gross understatement.

It is never easy to maintain a crew when working out of a camp environment such as these men were forced to do. Crew turnover rates were high in such situations, but nowhere near what he found when he returned. It seemed the only men who were remaining calm throughout this were those who had been hired from the Hoopa and Yurok Reservations nearby. The Indians just accepted what was and did their jobs. One of Mr. Wallace's first actions on his return was to get with his brother, Wilbur, also part of the crew, to find out what had been happening.

Wilbur spent hours bringing his brother up to date on the events that had been transpiring in his absence. He told him of the nightly visitations by their large friends... the myriad of tracks he'd left during these visits. Wilbur did his best to convince his brother that what he was relating had actually happened, but didn't appear to be having a lot of luck in doing so. Ray knew the situation they were in... It was fall and time for construction was running out faster than he wanted. They were under contract to finish this job and it was going to be difficult to do so with his crew deserting him over some fanciful stories.

Wilbur was not well pleased that his brother could not accept the facts as he understood them and as he had seen them unfold before his own eyes, so hoped to create more of an effect when he told him of the vandalism that had been perpetrated along with the tracks being left.

First, Wilbur explained, they came to work to find a nearly full fifty-five-gallon drum of diesel fuel used in the machinery to be missing. The tracks indicated that the being had picked up the barrel and carried it down the road, crossing to the open side before tossing it one hundred-seventy-five feet from the road in the heavy brush. He noted to his brother that that drum would weigh over four hundred pounds with as much fuel as it contained. Diesel fuel weighs, he reminded Ray, about seven and a half pounds per gallon... less than water, but still most substantial. When Ray wondered if it might have been rolled down the road grade rather than carried, Wilbur explained that there had been no tracks in the soft soil indicating it had been either dragged or rolled... and they would have certainly been evident had they been there. That barrel, he explained had been carried to the point where if could most effectively been thrown over the bank!

On another occasion, Wilbur explained, they had returned in the morning to find a tire off one of their earth mover scrapers, about two hundred fifty pounds in weight, had been carried and rolled for more than a quarter of a mile back toward the equipment dump at Louse Camp before being hurled, not rolled, into a very steep ravine... again, the creature had located the spot where it was most effectively released to create the maximum discomfort for those who would have to retrieve it from the canyon bottom. This fact was not well received by the senior Wallace brother and he asked Wilbur if this was all he had to share?

When Wilbur told him it was not, Ray resigned himself to hear more. Although he was not sure of the truth of what he was hearing, he could imagine the deleterious effects it would have on the morale of his crew. Even as Wilbur began his next account, Ray was thinking of some way to counter the negative effects this was having on his crew.

Most recently, Wilbur explained, Jerry Crew had returned to the dump at Louse Camp and found that a section of eighteen-inch steel culvert pipe was missing from the stack. This twenty-four-foot section of pipe would weigh approximately three hundred and sixty pounds. There were no machine tracks anywhere near that stack and it was clearly gone! It was later found at the bottom of a steep sided ravine approximately a hundred and fifty feet from the road way.

To say that Ray Wallace was not totally convinced would not be an exaggeration at this point, but he left his brother with a new respect for the events of the last six weeks. It was beyond his imagination how someone could have sneaked in and did all that was attributed to these mythical beings, for, after all, that's what they were, weren't they?

It was mark of the times in the 1940s and 1950s that on virtually all roads, any spring adjacent

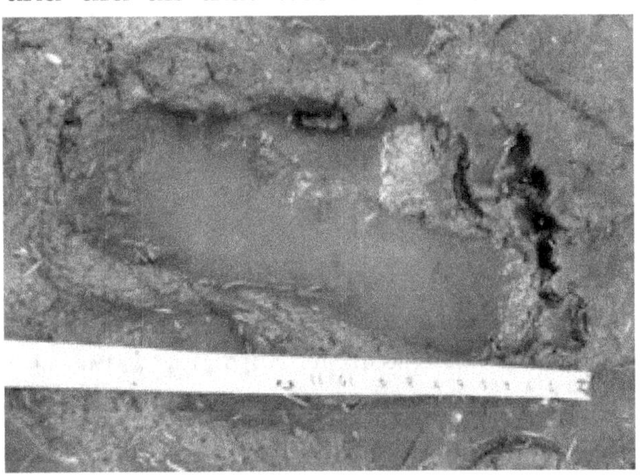

Print in Mud by Culvert

to the road was developed and a trough installed to provide a source of water to the drivers bringing their trucks in or out to fill their reserve tanks for their water-cooled brakes and to help motorists along their way should their engines overheat... a very common occurrence in cars of that era. A serendipitous side effect of this was a wonderful place to quench one's thirst with the finest water to be had anywhere and to refill, as needed, the desert bag that depended from the front bumper, cooling the water within through the action of evaporation. Today, that function has pretty much been replaced by Convenience Stores and bottled water that's marketed as "pure spring..." but actually comes out of a well and is, in fact the most polluted water one can purchase due to the greases falling into it from bottling machinery.   In the area we are discussing here one such spring tank was to be found not far from the supply dump at Louse Camp. It was here that Ray Wallace stopped on that day in week two of October and found, there in the mud around the tank overflow, masses of tracks... from, as Dr. Sanderson said, "Mr. Bigfoot".

At this point, the road builder knew that all that had been told him regarding the events of these past weeks were true. He could no longer remain skeptical or doubtful in the light of this new evidence he knew, because of the sequence of events of this day, could not have been faked or hoaxed for him to find. Mr. Bigfoot did exist... and he did exist on his job site... and he was a major obstacle in the way of completing this project on time.

About this time, another player came into the picture on this remote road site. Ray Kerr from Eureka had heard of the occurrences on this job and was most interested in tracking this creature. Since he was an experienced cat skinner, he was hired immediately with the understanding that he would work a full shift during the day and any tracking he did would be on his own time after his work for the day was complete. As this was Mr. Kerr's intent, he readily agreed. Ray brought with him, however, something more.

Along with Ray Kerr came Bob Breazeale, his wife Leslie, then age thirty-five years and Bob's four highly experienced hunting dogs. Bob came with a hunter's background. He and his dogs had hunted world-wide and his dogs were the envy of all who had seen them work. He carried a "large caliber, English made rifle" that seemed to be to the interest of all those in camp. Bob, Leslie and the four dogs were not hired on the road crew but came along simply to hunt and track. Their seeming goal was to capture or kill one of these "beasts" that were wreaking such havoc on the Wallace Brothers Construction Company's crew roster. The date this began is not certain, but on fifteen October, 1958, Ray Kerr and Leslie Breazeale were riding in Ray's vehicle when Ray spotted a "gigantic humanoid covered in six-inch-long dark brown hair" crouched by the side of the road.

According to their description of events as reported in a feature article in the San Francisco Chronicle on Tuesday, October 16th, 1958, Leslie Breazeale was dozing in the seat while Ray Kerr was driving. While Leslie did not see the creature init-ially, Ray did and he repor-ted that it rose from its crouched posi-tion to full height and in two steps cros-sed the twenty-foot-wide roadway and exited over the bank.

Hunting Hounds

The range from the vehicle to the being was approximately forty feet. This occurred within one and a half miles of the Company's supply dump at Louse Camp.

Leslie testified that the action of Ray braking precipitously woke her from her slumbers immediately and she witnessed the creature as he left the roadway after having crossed in front of their vehicle. Measurements were taken and recorded. The roadway measured twenty feet in width and the being was approximately forty feet in front of where the car was when the brakes were initially applied. The tracks were sixteen inches in length and were very much the same as a human track would be in that same soil, only much deeper and the stride very much longer.

Bob Breazeale was immediately notified of this occurrence and quickly put his dogs on the trail. He transported them to the scene and released all four of them together… they left the roadway in hot pursuit and were never seen again. It has been reported that the skin and bones of these dogs were found splattered on trees in the area sometime later, but that has never been confirmed.

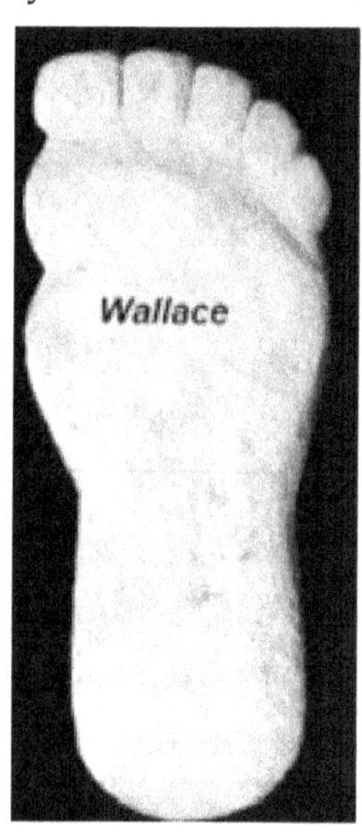

Print Comparison

With the rapid departure of his hunter and the vaunted dogs, Ray Wallace then resorted to a bit of subterfuge on his own part. Desperate to keep his crew intact, the contractor hit upon the idea of making an over-sized foot of wood and leaving tracks of his own making in an attempt to convince workers that it was simply he and not some mysterious, mythical creature leaving large tracks all over the mountains. It was a sad footnote on an otherwise stimulating series of events in a hidden mountain hideaway. Fortunately, the faked tracks are easily recognized and no tracker of any ability at all would ever mistake them for the real things. The first thing Mr. Wallace learned about making fake tracks with a sixteen inch long and six- or seven-inch-wide prosthesis is that, although his body weight did not change, his area on the ground did, creating the effect of wearing snowshoes. In essence, it was very difficult to leave prints in any but the softest of soils. He simply could not place enough pressure per square inch to penetrate the soil to any extent. This is a dynamic that has faced all who followed who would emulate his endeavor. Unfortunately, there are those who do not know better who continue, to this day, to ascribe to the "fact" that the Jerry Crew prints were made by the owner of the company (an owner who was out of the country at the time) to fool the people in the area.

Even at this early stage of reporting of this phenomenon, the doubters, nay-sayers and haters existed. Andrew Genzoli found out first-hand the results of speaking out on something so unknown as he suffered, in the words of Dr. Sanderson, "brickbats from deniers..." Sanderson went on to decry the attitudes of "educated men" who refused to even hear what was presented on the subject when approached with it. At least, however, that group remained quiet.

Dr. Grover Krantz, Professor of Anatomy and Paleontology, Washington State University, stated that, "If I had the body of a sasquatch in my office, I would have to physically drag those scientists into my office and rub their face on the body before they would admit to even having seen it..."

There is, however, one group that has no problem

speaking out loudly and in quantity... those who disagree with the idea. These people take many forms and guises. Some are zealously religious and are afraid this information might, somehow, undermine their faith. I have had conversations with this type of person and they seem of two general types.

There is first, the person who denies even the dinosaurs ever existed because, after all, "the earth is only five thousand years old because that is when the Bible starts." The second is a bit more pragmatic in that he does admit that dinosaurs were once around, they were just "pre-flood" and somehow (he doesn't explain exactly how) missed the boat. With this fellow, it's the timeline that trips him up.

Now, lest I be thought here to be anti-religious, let me state that I, myself, hold the Priesthood... but I can still think for myself on matters. I understand that there may be many things we don't yet know and act accordingly.

This brings up the third class of nay-sayer... the "pure science" group. If it is not discovered by "science" and cataloged by them, it cannot exist. These are those who state that observation is invalid... that eye-witness accounts are not acceptable. (I've always found it amazing that my eye-witness testimony can send a person to the death chamber, but it's not acceptable to these people as to the possible existence of a life form...)

It should be again be noted that proof is a very personal thing. I cannot prove anything to anyone... I can only offer evidence. It is up to the individual to decide whether that evidence is sufficient to constitute proof in his mind. It is incumbent on those desiring proof that they do more to obtain that proof than watch the latest trendy television show on the subject... if one desires proof, one must be willing to go forth and seek that proof. There is one other trait that is even more difficult for me to understand, and that is he who ranges from total understanding of the existence of this being to those with merely an understanding of that fact but who REFUSE to accept the premise that others may have observed some trait, facet or ability in it that they have not... and that they cannot accept for some reason. It amazes me that people can be so narrow minded as to insist that their understanding of the creature is the ONLY understanding to be heard!

I came across one fact early on... those who are the loudest in their denial of the existence of this being... those who are the loudest in attacking others are those with the least actual experience in the field. I never cease to be amazed by how vicious people can be to one another over something that has not harm in it... to the individual attacking, it really makes absolutely NO difference whether the creature we call bigfoot or sasquatch exists or not... for him, the sun will still rise tomorrow... his mortgage payment will still be due on the tenth of the month... his job will still be there on Monday morning, but he allows himself to be vitriolic in his attacks on others because they have experiences they interpret other than his own.

Test Photo 2

2.

A Complete Change of Thought

One of the most difficult aspects of reporting a sighting is the abject   rejection that one gets from so many people. As Stewart soon learned, being questioned by family and close friends… those who should stand by us and support us… was a major jolt, but that was only a primer for what happened when he shared it online. The rejection and ridicule were terrible. Even with photos to support the claims, the haters simply declared it all a hoax perpetrated to gain attention.

Fortunately, there were those wanting answers… those who appreciated the sharing of data and information. It was this encouragement that allowed the exploration to continue. Of course, in this situation, one becomes more discreet and much more careful where he shares his work and, fortunately, Stewart never lost his passion to continue.

Those positive people tend to increase our satisfaction and comfort and help us to go on seeking out conferences and meetings where we can learn from others' experiences.

At one such conference, Stewart met my friend and mentor from Oklahoma who is known for her keen knowledge about the sasquatch people and is fortunate enough to have a high level of activity on her home acreage. This ground was a heavily wooded, very isolated place in southeast Oklahoma, near Lake Eufaula.

This lady is of Cherokee descent, having a Cherokee Grandmother and has a keen interest in all things cultural coming from her people. To this end, she organized a Native American drum making workshop to take place on her property over a long weekend. A Salish woman from Washington, the state, not the city, was to be the instructor and there would be about a dozen people, most with Native American ancestry, in attendance.

Although Stewart was not particularly interested in making drums, he paid the fees just for the opportunity to be on the property and to, perhaps learn something more from the woman. This gracious woman gave him prior approval to scout around the property with the single caveat that he would not share the property address with any third parties.

The property was remote and heavily wooded, having rolling hills and creeks. The highway... the only public access to the area, was about a mile away. To get onto the property from a public road required traversing a private road winding through neighbor's lands... neighbors who were not the friendliest of people and who kept mostly to themselves. Every pickup came complete with a full gun rack to deal with varmints and any trespassers with an attitude. These were not people one would wish to impose upon under any circumstances.

Even though the hostess had advised these neighbors of the workshop, the visitors still got some very stern looks as they passed by on their way to the workshop venue… even those who had been invited in by their neighbor.

The hostess related several stories of incidents on her land. Many of these stories sounded quite incredible… too incredible, actually, to be believed easily. She had names for several of the Big Guys populating her land… there was Owl Caller, Indigo, Blue, Baby Girl and more. She spoke of having a hidden meeting place where she would leave regular gifts for them. She had a garden and orchard area specifically allotted for them and the standing rule was 50-50… they were entitled to half, but she needed half for her family.

When this habituation process is mentioned, many people simply will not accept this even though there are scores of documented stories of this by the various research organizations. "I must say," Stewart stated, "I was no different. This was not something I could accept. It sounded far to incredible and I assumed it was given more to embellishment than to fact." It seems we all tend to discount that which is out of our usual comfort zone or way of thinking.

Stewart relates, "On the first day of the workshop, I was away in the woods, sitting, standing and walking to various areas looking for any kind of possible contact. I heard a few distant knocks, a couple of calls and hoots, but nothing was close by."

It was his great desire to see one of these people up close and obtain a clear video or picture of them. He said, "Sadly, for my naïve way of thinking, this was their house and they weren't about to display themselves for me." There, are, unfortunately, too many people, even dedicated researchers who believe the sasquatch are nothing more than dumb animals that can be baited or lured like some kind of game animal. What they do not realize is that these people possess intellect and are quite capable of higher cognitive thinking.

Stewart stated, "With my error in thinking, I had spent the better part of one day and a night with little to show for my efforts. I felt, at times, like I was the one being manipulated and run around in circles with faint calls and movement noises. If there was a lot of activity around here, why wasn't I capturing it?"

There was a time, at the end of a dead-end road to the west of his hosts' property, feeling very frustrated with the lack of success, with no one else anywhere around, that that feeling of being watched returned again very suddenly as darkness approached. There HAD to be someone or something close by and watching. He had heard that one could talk to them and be understood, but was not something in which he had placed any faith. It was way too far out in the field for him, but discouragement had him to the point of being willing to try almost anything.

With nothing left to lose at this point, he began loudly, "Look, I'm not here to hurt you," feeling a bit stupid talking to the trees, at the end of a dirt road with no one around and just sitting on the tailgate of the truck. "I just want to see you clearly, up close… that's all. I'm not here to disturb you in any way… I just want to see you…" On and on his speech went, with him speaking sincerely to the woods.

There was no return sound save a very subtle movement off to his right side. The feeling of having been heard by something or someone was verified when another of the attendees came down the road to check. This fellow was also in Law Enforcement, was of Choctaw lineage and a friend who had taught much about the bigfoot people. He said, "Do you know something big just moved away as I came down the road?"

The following morning, Stewart again hiked into the woods, lurking about with his video camera and digital recorder. He then returned to the group for lunch and just relaxed and rested for a bit. About an hour before the afternoon session was scheduled to begin, three ladies who had been on a walk came rushing back into camp to state that they had just seen a bigfoot peeking at them! They were extremely excited, but most of the group could not generate enough interest to even wish to see the photo they had taken of their sighting. Stewart was highly interested but disturbed by this event. He felt shorted for he had spent nearly two days deliberately trying to see one without success and these ladies just casually hiked out and ran into one!

One lady produced a photograph that showed her subject with his head peering out from behind a tree! That was even more frustrating since she had gotten a picture of him! And had done so in a thirty-minute walk from the workshop! After discussing their sighting, he talked the two of them into showing him the spot they had sighted him before the afternoon session began. Although it was only about a quarter mile, it required some hiking through wooded terrain and on the way across an open pasture to access the woods, they observed a footprint in the soil. The depth supported the fact that this was something much heavier than the three of them as their feet left no lasting impressions in the pasture. The toes were well defined with a distortion that indicated whatever had made that print had made an alteration in stride or direction at that point. The track was not deteriorated, but was sharp, supporting the supposition that it was very recent. There were other, less well-defined tracks there, but nothing as clear and distinct as the first one found.

Travel here was through wooded forest to a very wide creek. There were no access roads and to reach this point meant a hike through moderately rugged terrain. The only other possible access would be through fenced off property belonging to a neighbor… probably neither a safe nor easy thing to accomplish. In summation, just say that this location is not a place where people commonly come to just hang out. It would take a very deliberate effort to reach this spot.

"We stopped on a rock ledge where the ladies had sat down to rest and enjoy the view. I could see the large pine tree about seventy-five yards down the hill where the subject was observed watching the ladies. He was side-peeking from around the base of the tall pine tree. I began scanning the base of the tree and surrounding woods. We didn't talk much but listened and soon began to hear clear, very distinctive footsteps in the dried leaves... The direction of travel seemed to be edging closer to us."

Stewart then spent nearly an hour videoing the woods from this point on the ledge. Briefly, at one point, he spotted a being moving about among the blocking limbs and leaves. The range was estimated, at this point, to be just a bit beyond a hundred yards and the being briefly moved to another position when he saw him... then he was gone... a fleeting moment, caught, at last, on video. From the time of first sighting until the subject was gone did not exceed twenty seconds. His movements were fast and he was acutely aware of their position. The being did not appear to be a human, certainly, but had a distinct foreign look to it.

Of course, the man was excited with the thought of getting something on video, even if it was not the clear closeup view he had wanted, but time was running out. This evening would be the  final evening of the workshop.

A fireside gathering was planned with native drum playing and other, similar, customs. He participated in the opening ceremonies then quietly slipped off to the west and into the woods. After about fifty yards, he turned south, doubled back and pointed his way east so as to not be easily tracked so as to not be pranked.

"I scanned my perimeter and quietly listened with each change of direction while using my night vision scope… I made my way back east near a break in the forest by the land owner's barn."

For almost an hour he watched and listened to the drum beats and the chanting of the songs. Common knowledge is that the bigfoot people show great interest in music with a gathering like this.

"I kept watch of the tree line… My heart nearly jumped up into my throat when I suddenly noticed a figure ease out of the woods about a hundred and fifty yards away. I began videoing, hoping I had just obtained a night vision video of this elusive subject. All my hopes faded as I realized it was the land owner… She didn't head back in the direction of the group, but rather, walked directly towards me! The closer she came, the more perplexed I was… could it be pure chance she was heading this way, I thought."

Eventually, she made her way to where Stewart was hiding and as he stood up, he asked, "How did you know I was here?"

Her response just about blew him away when she said, "They told me you were here… Do you want to go see them?"

At this point, the man realized that this was far beyond pure chance. This was far beyond mere luck, but still his mind refused to fully accept it as real. It was like something merely dreamed, but he gamely answered, "Yes, let's go."

She instructed him to use no lights or cameras without asking her first as she related that the native people do not like these items or anything that is held up to the face.

The two made a brief stop by the fireside and ascertained that all were still there. They then walked a hundred or so yards to the north pasture and stopped by a wooden cattle fence and gate. There was about a two-thirds moon shining brightly and vision in the open was no problem although the shaded areas under the thick canopy was totally black. As the pair leaned against the gate, Arla began to call out names of those who lived near her. Indigo… Owl Caller… Mau Mau… Baby Sister… The brush and woods were cut back about ten feet beyond the fence in this area, affording a nice range of vision in the bright moonlight.

Very soon there were the sounds of heavy footfalls coming up from the creek bottom. They made whatever was making them sound very big with the snapping and crunching through the heavy brush. More footfalls began from the northeast in a very similar way and then, more from the southeast. It sounded very much like an army of giants making their way through the woods. Stewart thought to himself, wow, this sounds very real and anxiety began to set in with him.

Carefully, he watched through the gaps in the leaves and branches for any positive sign of lights that might have been being used for those making their way through that thick stuff. There were none to be seen… he thought… It is not a good idea walking in the Oklahoma woods at night without lights… This close to water, and with the temperatures warming up with spring, the Copperheads and Water Moccasins could well be out hunting anywhere around.

For the next part of this saga, we are going to listen to Stewarts own words to describe the activity…

"As the sounds drew closer, I was getting apprehensive. Whatever was out there was big, as I could feel some of the foot plants shake the ground beneath my feet. About ten feet back beyond the clear cut, all walking sounds came to a halt. I honestly was fearful now. Intense emotions began coursing through me! Whatever was over there sounded big. Really big… And they were just ten to fifteen feet from me… not one, but several."

"'Can I turn on my night vision?' I nervously asked my host. She gave me an odd smile and replied 'Sure', as if knowing something I didn't. I pressed the 'ON' switch and absolutely nothing happened. I pressed it again and still nothing. All my suspicions of this possibly being a prank vanished in seconds. I knew there was no way anyone could affect my brand-new scope… NO WAY… Yes, it wasn't working, and I had used it only ten minutes before. I tried the switch again and nothing…"

"I remembered reading one story where some researcher's electronics went on the blink when the bigfoot subjects came near their location. I didn't believe it, but now I was experiencing the very same thing. I had no logical answer for this. My apprehensions became stronger. Fear of the unknown or something we don't understand is just human nature."

"'It won't work,' I exclaimed.

The host calmly smiled back at me and said, 'Try it now'. I pressed the switch and the scope came right on without a problem."

Stewart was amazed by the events unfolding here and was more than a bit disconcerted by this time, as one can imagine. He thought, almost out loud… Is she some kind of witch?

It was at this time that it dawned on him that it could be those beings in the brush. Just the thought of this was nearly terrifying to him. He tried to see through the brush but could see absolutely nothing… it was just too thick and all light was blocked out… even when using the Infrared fill on the scope. All that was at all visible were the leaves and the limbs but nothing beyond that. At that point, he gave up on the scope, turned it off and holstered it, saying, "I can't see anything."

"Do you want to go inside there and see them," his host asked while opening the gate.

"I can truthfully say, at this point, I was terrified, but my testosterone pride would not let me wimp out in front of a woman. If she had been a man, I would have said, 'That's ok, I've seen enough'," Stewart admitted.

Suddenly, for him, entering that gate and coming so close to something so large and so powerful had lost a degree of importance and was not a priority with this feeling of vulnerability taking control… a feeling very foreign in this man. However, with Arla standing there calmly, he answered with a very sheepish, "Okay…" and reluctantly followed her through the gate.

They turned left into a narrow, cut-over trail through the brush and he could see her well enough to follow while in the open pasture, but turning into the trees made what light there was, completely disappear. He closed his following distance to the point he could about touch her and kept her dim outline firmly in sight. He was, he said, "Afraid to look anywhere else."

There was heavy movement in the brush on both sides but he refused to even look. For some reason, he didn't even feel he had the strength to turn his head and look, expecting at any moment for something to jump out and grab him. It was, he said, "Like riding a scary ride at Disneyland, only multiplied about ten-fold." It was terrifying to be in their house and virtually blind. "It was not a comfort zone for me as I silently prayed for God to be with me and divinely protect me."

Stewart trailed the host closely knowing she was their friend and was not about to fall behind. He was holding his camcorder while nervously fidgeting with it just to occupy his hands to keep them from shaking with fear. One thing certain, he was not going to put that camera up to his face and anger whoever that was in the brush so nearby.

People in the public are often critical of people who do not get pictures of a visual sighting, but, as Stewart said, "They cannot begin to comprehend how intense it is to get close to these subjects. Sitting safely behind their computer keyboards is not anything like the real deal. It is intense beyond words."

As he walked, his fidgeting caused the camera to take several photos of leaves and trees as well as the underbrush, and on one occasion, totally by accident, a Big Guy peeking out from the thick vegetation beside the trail. The picture is not in telephoto mode, it is the actual distance he was.

The two moved in about a hundred yards onto a trail that intersected at right angles. Following this second trail led to a clearing about thirty feet in diameter where there were two folding chairs residing in the middle with gifting bags hanging from the trees about six to eight feet off the ground. It was the host's habit to leave snacks for them suspended high off the ground, in order to prevent pilfering by rodents and other denizens of these environs but still within easy reach of the sasquatch.

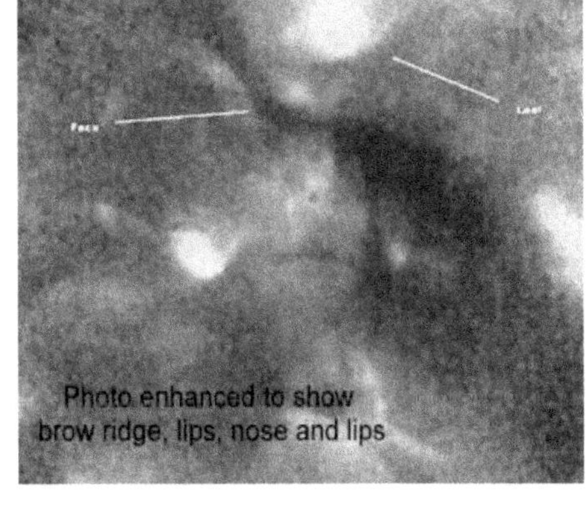

A break in the canopy here allowed a filtered moonlight in and it was possible to see, but he was still not able to muster the courage to take a good look around the area. He watched as Arla began calling and talking to them... then it happened...

"A crystal-clear voice spoke in my head. It was not some passing thought, but a clear and very direct communication, asking me, 'Why are you carrying a gun?'"

"It shocked me hearing this voice in my mind and I felt 'Man, am I losing it here?' not believing what was unfolding. It was like the gain in my head was turned up ten-fold and the voice asked again, 'WHY ARE YOU CARRYING A GUN?' I felt like the scare crow in the 'Wizard of Oz' hearing that booming voice… terrified, but I instantly knew they were speaking to me."

"Being retired law-enforcement, I often legally carry a well-hidden pistol on my person, like a set of keys. People will never see it discreetly tucked away, but if I must defend myself or another, it is there solely for protection. That habit saved my life once while scouting out some white bass fishing spots along a remote creek. Three feral pit bulls attempted to attack me. If I hadn't had one of my .45 ACP pistols with me, I would not be (relating) this story now."

"No one in the entire group, other than another active-duty law-enforcement attendee knew I had a weapon. That told me whatever these subjects were, they possessed a keen, intuitive sense beyond the normal… one more thing telling me these subjects were not what I expected."

"I don't know how I knew how to answer them in my head. Maybe it was a primal ability we all had at one point in time, but our civilized ways of living insulated from the country caused it to diminish over time."

"In my head, I replied, 'It is a tool of protection that I have carried on my job… I have carried it around my close friends and family… I have never hurt them with it and I will not hurt you now.'"

"The voice then replied, 'Step away from (her)'"

"I turned and spoke to her, 'They are telling me to move away from you!'

Instead of being shocked and surprised, she just warmly smiled back and said, 'Well, do what they tell you to do.' She wasn't one bit surprised."

"I stood there for about thirty seconds, pondering my fate. 'Were they going to attack me and wanted their friend out of the way?' I had no clear idea what was right, but my adrenalin was flowing. I turned back towards the main trail and started walking slowly. I felt like I was stepping off into the Grand Canyon of uncertainty. It is probably one of the hardest things I have ever done. Terrified and praying, walking  back to the trail intersection. I paused and looked back the way we came in, then turned to my left where it led deeper into the woods. There were ribbons of moonlight in various areas giving the dark woods a surrealistic look. About fifty feet down the trail, it was illuminated by a wide beam of moonlight filtering through the tree canopy. Very slowly, a hulking ten-foot (tall) figure stepped out from behind a large pine tree beside the trail into this beam of light. It was just like Ed Sullivan used to do, stepping out from the curtain into the spotlight some fifty years ago. It seemed almost choreographed."

"The figure was an absolute replica of the comparison charts you see of a Bigfoot standing next to a normal sized man. He had huge, wide shoulders and a conical shaped head and (a) thick brow ridge. It was obvious that I could see he was male and probably eight-hundred to a thousand pounds in weight… a sheer monster in size that I wouldn't dare make angry. I couldn't make out his eyes in detail, but his face was staring right at me and he was slowly swaying from side to side, standing in the middle of the trail."

"'Can you see him?' my host asked."

"'Yes, I can see him,' I answered, 'He's swaying.'"

"I do not know the significance of this motion, but my Host replied, 'Sway with him.'"

"I matched his sway motion like a tree moving in a soft wind. A stirring then began in the brush off to my right side. A figure rose up from the chest-level brush right in the middle of another ribbon of moonlight. He looked like a younger male standing about fifteen to twenty feet away. He looked like something right out of a Stephen Spielberg movie. He was about six feet in height with a much larger frame. My jaw dropped open in astonishment as I looked right at him. This was the close, clear look I had been wanting. I could easily make out his facial expressions gawking back at me. He was just as fascinated with me as I was getting a close look at him."

"At another point right in front of me, another subject stepped out onto the trail and it was obvious she was female. They both were about my height and I realized these were not grown adults but two juveniles. Behind the young female and in front of the big male, an eight-foot-tall subject stood out which I assumed to be the mother. They all held positions in areas of moonlight allowing me to get a good, clear look at them… Exactly what I had been seeking since I got there!"

"As a parent, it immediately struck me like a ton of bricks… They were trusting me enough to allow their young to come close and get a look at me. This spoke volumes to me and my emotions took a drastic U-turn. Instead of fear, I was now flooded with elation and amazement at this once-in-a-lifetime experience! They had accepted me!"

"My mind relaxed but it was to be very short lived. I then noticed a small, three-foot tall figure toddling around in front of the female standing on the trail in front of me. I couldn't clearly see it because of the tall brush shielding off the moonlight, but it moved just like a young toddler would do. It stopped and looked up, right at me, the distance being about fifteen feet away. All at once, it quickly shuffled towards me with its little arms reaching up like my granddaughters do when they want me to pick them up. I nearly messed my pants at this unexpected event. Having these subjects close to me was incredibly intense, but the thought of a little, hairy one latching on to me was more than my blood pressure could take!"

"'What if I startle it? What if it cries out and Big Daddy thinks I hurt it?' came into my mind. 'The Big Guy might come up here and tear me in half!' I think my heart skipped a couple of beats before the young female quickly moved forward and grabbed the baby and pulled it back to her by one arm. I started breathing again."

"'I'm good, let's go!' I called out to my Host, still concerned the little guy might get away and make another dash towards me! I wanted to leave now. Things were getting to much for me to take. I wasn't scared of the little baby. I just wasn't sure it knew who I was. I didn't want to give it a chance to cry out near me."

"As we were backing out of there, the big male started letting out a perfect owl call. It sounded very authentic, but the resonance sounded like the owl was eight-hundred pounds… Way too full for the smaller bird."

Stewart was able to get an excellent audio recording of the owl call. At no time during this encounter, did he ever intentionally point a camera at them… he said he was actually afraid to do so. Even so, he walked out of that encounter emotionally spent. He had traversed from one end of the emotional spectrum to the other in about a single hour. He said… "While it had seemed to me to have all happened in about fifteen minutes, the others assured me it was very nearly an hour."

Stewart stated, "I knew this was a once in a lifetime experience that few might believe but I know it happened and I had the land-owner with me as a witness. The police officer friend also heard the owl calls as we were leaving."

Just to have the entire incident recorded officially, he wrote out a sworn statement and sent it to the North American Bigfoot Search Team. It is listed in their sightings as caption story #63.

That evening, Stewart's entire perspective changed concerning just who and what they are and the amazing abilities they possess. From someone who did not believe that telepathy could exist as a form of communication, he realized they are a tribe of PEOPLE… not some dumb animals… with family groups and clans. Most of society does not believe they exist, which is a good thing for the survival of these beautiful beings. There are many research groups who still regard them as mere animals to be hunted like some trophy.

"My hope and prayer is that the Forest People will escape the fate of other indigenous peoples. Though very unlike us in some ways, (and) very much like us in others, they are truly, 'America's last FREE tribe!'"

Richard Stewart Taylor

Chapter 3
The Patterson - Gimlin Film

Bill Munns is an artist... he created monsters for the movies and he ran the school teaching others that art! He wrote this book in 2014 and I found the dedication so profound that I published it here... with Bill's permission, course... I hope you will enjoy it and learn from it... and perhaps gain the desire to read more!   Thom Cantrall

When Roger Met Patty
Dedication

This book is dedicated, first and foremost, to the actual participants of the Patterson-Gimlin Film Mystery (one

deceased and one still living) because they have seen

Bill Munns at work

their lives transformed (or cursed, in some cases) by this film. It has made them famous and forced them to adjust their lives to the controversy of the film. In a sense, they have lost a significant portion of their lives to this film, so powerful in its impact, and so turbulent its controversies.

It defined Roger Patterson's life, literally becoming his legacy, but it also put his life under a public microscope, and as his life was not that of a "regular guy" with a steady

job, every arguable flaw or mistake, every irresponsible or thoughtless action he may have done, is documented, debated and publicly examined. People intent on showing the film to be a hoax have vilified Roger with an intensity that still astonishes me. So first and foremost, this book is dedicated to Roger, because he did not hoax this film. Whatever else in his life you may wish to criticize him for, as being flawed or fake, this film is honest and what he claimed to film was true.

Second in dedication is Bob Gimlin. When he agreed to accompany Roger on this expedition to search for Bigfoot evidence, he could not possibly have known that the encounter in the Bluff Creek forest would devour his life like the shark in JAWS. This film would make him world famous; it would make total strangers hate him, it would provoke people to impugn his personal integrity again and again, and it would be something he could never erase, never ignore, never resolve. People examining his life obsessed with second-guessing his decisions, his actions, his words and his demeanor. But those who do are fools, because they judge Bob without full knowledge of his thoughts, his personal decisions and private considerations. He has attempted to maintain some semblance of privacy in his life, despite being something of a "public figure" by virtue of this film, and his public actions and statements are driven by choices, considerations and factors in his private life, which he has every right to try and keep private. So, to judge Bob is to indulge in petty character assassination, because to defend himself, he would need to surrender his privacy, and every human needs some privacy just to be human.

In October, 1967, something happened to Bob that was so rare, so bizarre, and so powerful in changing the course of his life, that we can never judge how well he has managed to endure this experience. How well would we, if fate put us there instead of him? But, while Roger passed away less than five years after the filming, Bob has endured 60 years, as of the writing of this book, of this film's grip on his life.

And yet the film has defied resolution during those years, holding his life, his reputation, in suspense. So, with this dedication, I hope that it gives Bob some comfort and some resolution or closure to the controversy. He saw something real that day, 60 years ago. It was not a guy in a fur costume. He's always known that. Hopefully, now more people can share what he has always known.

Third in dedication is Patricia Patterson, Roger's widow. When Roger packed up the truck to go with Bob Gimlin in early October, 1967, headed for Bluff Creek, CA, Patricia was a housewife with three young kids and no financial stability in their life. Her problems were the problems of millions of women, to keep a household going and raise some kids despite financial insecurity. She also was struggling with the difficulty of seeing her husband endure the crude radiation treatments of the era for Roger's cancer (Hodgkin's Lymphoma). Then he came home from his trip and announced to the world what he had filmed. And Patricia's life went from being challenging and understandable to bizarre and unprecedented.

Mothers the world over deal with comforting their children who may be bullied or mocked at school, for being too fat,

too skinny, too dumb, too smart, or too unattractive. But Patricia had to comfort her kids who were bullied or mocked for having a father who claimed to have filmed "Bigfoot". She watched the media circus Roger engaged in with the assistance of his brother-in-law, Al DeAtley as they four-walled the movie around the country. She

Patty Frame 72

watched his cancer return and lead to his death in 1972. she found herself alone, with three kids, no estate or assets, except a bizarre piece of film that TV producers would pay good money for rights to broadcast. How does a woman deal with such a situation? What friend do you ask, what course do you take, what professional do you hire, to help you make the kind of decisions she was now faced with, trying to manage this one utterly unique and controversial asset?

Since Roger's passing, 50 years ago, Patricia has been approached by people who try to prove the film is real, others who are determined to prove it a hoax, and media producers who want to license and exploit the film for program content. She has been poorly served by some of these people, and outright deceived by a few. Now, her own health is declining and she deserves some closure on this film which looms over her life. So, this book is dedicated as well to her life, her challenges, and hopefully, some closure to this film controversy that has caused he so much grief and aggravation…

Bill Munns

http://www.amazon.com/William-Munns-When-Roger-Patty/dp/B00N4F299S/ref=sr_1_2?s=books&ie=UTF8&qid=1409518989&sr=1-2&keywords=when+roger+met+patty

Letter of permission to use this portion…

Thom:

Thank you for your letter. Yes, you most certainly may share the dedication on any groups you participate in.

It was a pleasure, indeed an honor, for me to write that book, and give some measure of vindication to Bob's story. He deserves such… Best regards, Bill

Five frames from the Patterson-Gimlin
Film Showing Patty's Walk as She Leaves
the Film Scene

Prelude to a Film

During the mid-sixties, two persons of note set into motion a set of circumstances that would culminate in the events that inspired this missive.

In 1958, two amateur rough stock rodeo riders happened to meet. Since they both rode bulls and broncs and hailed from the same area of the country, they had a commonality that brought them together and caused a friendship of sorts to grow between them.

Bob Gimlin was born in 1931 in a small town near Branson, Missouri. Bob was of Native

Bull Rider

American descent, being Chiricahua Apache Indian. Bob, with his family, moved to central Washington about 1940 when he was nine years old. The family settled in Wapato… in the valley very near where Bob resides today. Their ranch was in the area of the vast Yakama Nation Reservation and a huge wild horse herd roamed the sage covered hills between the ranch and the reservation. Bob, as a very young person made considerable money for the time catching and rough breaking these mustangs before offering them for sale at the local auctions. I suggest that anyone who has the opportunity inveigle Bob into telling you his story of how he would catch this wild stock, hook them to a no-longer-used wagon and go tearing across the sage covered wilderness until, often, he ended up with only the tongue and maybe the front axle remaining of the wagon!

After serving in the U.S. Navy during the Korean Conflict

from 1950 through 1953 and as a result of this ability to handle rough stock, Bob began a short career in rodeo. It was during this time that he met Roger Patterson.

Roger was a couple of years younger than Bob, being born on Valentine's Day, 1933. He originally hailed from Wall, South Dakota where, legend has it, it is illegal for any male over age of three to appear in public without a cowboy hat and (optional) chaps. Cowboy boots were also required prior to actually walking! Naturally, Roger learned to rodeo as well, and being a smaller built person with a tremendous natural athletic ability, was destined for rough stock work.

As Bob relates the story, he had retired from full time rodeo riding by the end of 1958 and returned to his Yakima, Washington area home to take a more "normal" job. He worked a lot as a truck driver and his bride, Judi worked in a local bank. Bob also kept an active business going on the side breaking and training young horses.

He stated that he was in town one day in the early 1960s, I believe, when he heard a voice call out his name and, turning, saw Roger Patterson coming towards him. Bob was excited to see his old friend and asked how he had been doing whereupon Roger explained that he had been ill... something called Hodgkin's disease (a form of Cancer) ... but, although he was ill and weak, he seemed to be in remission at the time. Conversation revealed Bob's work with the young stock and when Roger questioned him about this, it turned out that Bob often took the colts right past Roger's place to exercise them and work them on the trails into the Cascade Mountains. Roger asked if it would be possible to just stop by his place and he'd load his pony in with Bob's and they could ride together.

As it turned out, this happened often in those years. And, perhaps prophetically, Roger had developed an interest in the subject of sasquatch... he would bring along a tape recorder and play tapes from people who had had encounters for Bob to listen to in the evening around the fire. Eventually, this expanded further to the two of them riding to likely areas while Roger searched for signs of this elusive

Mount St. Helens

hominid.

One of their favorite places was the area around Mount Saint Helens in western Washington. There were many roads and trails in the area and it was a very beautiful place to ride. Bob is quick to tell us he never saw anything here to change his status as a skeptic. In all their wandering, he never found so much as a footprint that he could believe was from a sasquatch.

Bob describes Roger as a "tinkerer" ... if he had something mechanical, he would figure out a way to make it better and attempt to change it... often successfully. Roger was also a saddle maker of no small skill.

Roger has often been described by critics as "never working". While it is unsure exactly what that means, it is true that he did not live a "standard" lifestyle, but it would seem that living with cancer would probably be enough to cause one to find alternative ways to make a living.

In late August of 1967, Roger received word that tracks of the sasquatch had been discovered in a remote area near Mount Saint Helens in southwestern Washington and he recruited Bob to carry their horses to the area and they could spend some time riding there to see what they could find. To that end, the pair drove the short distance from their homes to the Saint Helens area with the idea of spending the Labor Day weekend, 1967, investigating the possibility of such a being living there.

The pair, with their stock, were greeted by incessant rain... and one has to experience a storm on the west slope of the Cascade Mountains to truly understand how wet, cold and miserable one can be in a very short time. Eventually, the pair gave up on their search as they could not get off the main roads, even with the horses due to conditions and they finally yielded to common sense and retreated back to the warmer, drier climes of eastern Washington.

Bob related that he had only been back at work for a matter of a few days when Roger stopped by his home to ask if he could drive them with their horses to California. It seems some tracks had been found in a place called Bluff Creek near an unimproved campground called Louse Camp. As will be recalled, this is the same Louse Camp used by the Ray Wallace construction crew as their main equipment dump in 1958.

Bob, who was working for a roofing company in Yakima, Washington as a hot-melt roofer, told Roger that it would be impossible for him to get away just then. He explained that he had stock to tend on the home ranch as well as he had a need to work as much as possible as the roofing season was dying down and they would soon be out of work entirely.

During the last week of September Bob told Roger that it appeared he would be available to make the trip by the end of the month. To this end, Roger put together everything needed to make the trip while Bob continued to work. This included borrowing one horse from a neighbor named Bob Heronimous. Chico was a good, stout roping horse, according to Bob Gimlin and could also be counted on to pack. Both Roger and Bob knew this horse well as Heronimous had used him in some of the work done for Roger prior to this trip.

When all was in readiness, the two men set out on the more than six-hundred-mile drive from Bob's home to Louse Camp in Northern California, arriving on either 30 September or 1 October 1967.

Bluff Creek Odyssey

California's fall rains, so prevalent in the north part of the state, greeted the two men as they made their preparations for their explorations and discoveries. Unfortunately, this rain destroyed the tracks that had lured them to this area. While they were discernible if one knew what they had been, to Bob, they were merely amorphous blobs in the mud.

This began a routine that was to continue for their entire stay... On the horses riding the back country during daylight hours, then in Bob's one ton truck, driving the roads at night in hopes of meeting one of the denizens of this area.

The pace was grueling and fatigue was a factor. The two men persevered day after day in their quest... Onion

Mountain, Blue Creek Mountain, Forest Road 15... Forest Road 12... if there was a road, they traveled it... Turtle Rock, Nikowitz Peak and the Siskyou Range were their home for this period.

Sightings were common of the animals that resided there... Bears, Cougars and deer were seen regularly. Bobcats, skunks and the myriad small critters were everywhere but there was not one instance of anything that could have been Sasquatch to be found in all their perambulations.

Trips were made the thirty miles or so to town for gas irregularly. They had spare cans they filled to minimize the number of trips they had to make as this was a monumental waste of time for them.

For twenty days, this routine prevailed... ride and drive... search and hope until, on the evening of the twentieth day, Bob told Roger that it was time to point back north. He explained that he had things to get done before winter set in and it was time to be about them. It was a Thursday evening when this occurred and as much as Roger tried to change Bob's mind, it was not going to be... Bob was headed home after the weekend.

"Okay," Roger commented, "what would you think about leaving me here and coming back to get me after a while?"

"Roger," Bob replied, "that is not going to happen. If I leave you here, you are going to be here until spring because I am not coming back!"

"Well, since we only have a few more days, how about we take a pack horse and go up into that good habitat area we saw the other day?" Roger continued, "We can stay overnight there a couple of nights and be back in time to load up and get out of here by the end of the weekend."

The agreement made, the pair made ready for a morning departure but Roger, not being ever mistaken as an "early riser" was a bit behind time when they finally left their base camp at Louse Camp. It was the 21st Day... a day of Destiny for these two men and the world at large.

The Encounter

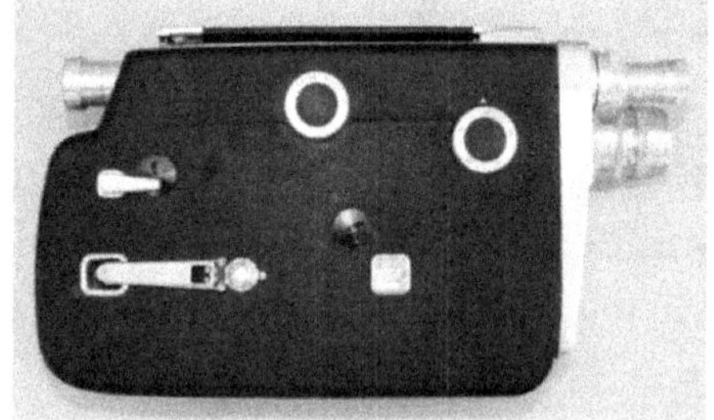

"What did you get on the film?" the man holding the rifle asked of the cameraman. "It looked good from here."

"I really don't know if I got anything," a discouraged Roger Patterson answered as he looked toward the western ridge to judge the time. "I ran out of film and I had to chase it clear across that flat. I fell down coming out of the crick, so I don't know if the danged camera even worked after that."

"I saw you fall," Bob answered. "I think I'm going to ride up the crick a ways and see if I can see where it went."

"No, Bob, please don't leave me alone. We don't have any idea if those other two are still here of if they have gone away. They don't seem to be dangerous... that one certainly wasn't at any rate."

"Stay here with me," Roger continued, "and let's see if we can make some casts of some of these tracks."

The two men did that. In the afternoon sun, they mixed plaster and poured it into the deep impressions left in the sandy soil of that bar alongside Bluff Creek that Friday in October so many years ago. While the plaster was drying,

Bob compared the tracks of his horse to those of the creature they had just watched walk away from them so nonchalantly a few minutes prior. It amazed him that his horse who weighed more than a half ton did not leave impressions nearly as deeply as had that creature they had just watched. Bob even tried jumping from a stump with his high-heeled cowboy boots on to see if he could simulate in any way, the depth of track the hairy beast had made. Try as he might, Bob's best efforts were dismal when compared with those left in the soil. He walked across the sandbar following the track-way and was amazed by the fact that, as casually as that thing had strolled from their view, the stride nearly doubled when it was no longer in their sight. He followed the tracks as far as the creek and found no more than wet spots on the rocks of the stream, apparently indicating where she had crossed the stream and climbed the steep ridge opposite them.

Frame 354 of the Film

It was the work of some hours to complete their tasks here this day and retreat towards their soogan at Louse Camp at the mouth of Notice Creek on the newly cut road. This camp had been their home for three solid weeks now as they had determined that this was the base of operations that would provide them the best chance of success in their venture... The sun was well down in this late October sky, nearly four pm, actually, by the time the two men reached their camp. Darkness was near by the time the saddle horses and pack horse had been stripped, wiped down and made ready for the night. The two men then loaded into Bob's one ton dually Chevy truck and headed into town.

Most of the thirty plus miles from Louse Camp out over Onion Mountain and down to Highway 96 that took them south toward the town of Willow Creek was done in the gloaming but it was full dark when they finally arrived in the town. That it was the first time that either of the men had been out of the mountains for any reason, other than for gas, in the past three weeks was noticed by no one. Roger and Bob looked up the man that had given them the clue that had led them into these mountains those weeks before.

In early September, Al Hodgson had sent word to Roger Patterson at his Yakima, Washington home... a track-way that had been found in the area. Roger immediately went to his saddle partner, Bob Gimlin to try to enlist him in quest for film sufficient to make a documentary on the legendary Sasquatch.

On that fateful Friday afternoon of 20 October 1967, about three- and one-half miles from the bridge over Bluff Creek at

Fig 2-7 Al Hodgson

Louse Camp, these two men rode out of myth and into Destiny. Bill Munns has described it as the minute of film that has sparked fifty years of controversy.

Following is the story the two men related to Al Hodgson that evening.

It was early afternoon before the riders reached their

destiny... Roger was in the lead followed closely by Bob about a horse length behind and him leading their pack horse with their gear. As they advanced up the sand and gravel banks left behind after the 1964 northern California floods, they approached a log laying in such a way that the upraised root wad had caused the stream to reroute around it, causing an unusual bend in the creek with a berm of sand built up on the opposite side from the fallen giant.

As first Roger, then Bob cleared the exposed roots, there stood, directly in front of them at a range of only a few yards a large, hair covered, bipedal primate more than seven feet tall and weighing in excess of six hundred and thirty pounds! When the two men first saw her, she was standing and

Bluff Creek Showing Creature,
LogJam and Bank Where Roger Tripped

merely watching their approach, making no real effort to escape.

Immediately, Roger grabbed for his hand wound movie camera. His horse, however, had a totally different idea of how this scenario should go down and immediately went into a wild fit of bucking and pitching. Roger was doing his best to clear his mount with the Kodak camera as his colt went to the ground, feet flailing. Being the athlete he was, Roger escaped his dilemma and ran for the edge of the creek and the waiting sasquatch.

Bob's horse was a bit more stable, though it took both hands to control the jittery, nervous mount. To better facilitate this, Bob and his pack horse parted ways, allowing the man to concentrate on what was happening with his pony and the creature in front of him. He watched intently as Roger started his camera while moving toward the creature.

In crossing the creek, Roger's attention was, of course, riveted on the being we have come to know as "Patty", so named by the Russian Humanoid expert, Dimitri Bayonov... In his attempt to negotiate the sand berm beside the stream, Roger tripped and fell on his forearms and knees. It took him only moments to regain his standing position but, by this time, Patty was moving slowly away from the scene of the initial encounter. Quickly, Roger moved across the gravel bar to a position where he could see the large being more clearly and without interference between them. When he achieved a spot that seemed to work for him, he called to Bob, still across the stream, "Cover me!"

On hearing this request, Bob rode his horse across the creek to a spot very near Roger, pulled his 30'06 rifle from its scabbard on the saddle and dismounted the horse to be more readily able to defend them should this being become dangerous. Further, since they had initially been drawn here by the discovery of three sets of tracks of different sizes and seemingly from different family members, it was feared that the other two may still be in the area.

It was when Bob dismounted with his rifle that the large creature turned and looked back at him. It is this famous turn that has been immortalized in the widely known but misnamed, "Frame 352." In actuality, it's Frame 354, but a miscount early on has mislabeled it for all these years! Once Bob had dismounted with his weapon, he simply stood and watched her walk away as Roger did his very best to get film of the enigma...

When she had finally disappeared from sight and Roger had walked back to Bob's side, Bob suggested that he ride after her to see where she went but Roger was still spooked by the possibility of there being two more if these individuals as well as the fact that it was late enough in the day by now to make it critical that they retrieve their missing stock, return to camp for the gear and supplies they'd need to cast some of the tracks left in the sand of this bar.

Roger reloaded his camera with a fresh roll and, as the pair worked the rest of the daylight away, he recorded the scenes generated in the process...

In Bob's Own Words

Following is a description in Bob's own words of the events of that day, October twentieth, 1967... The Day of Destiny... No corrections have been made to his words, it is as he said it, carefully transcribed from an oral presentation Bob did a very few years ago.

"...I left out early. I always got up early... Roger didn't. He usually slept in a little more than me. I'm kind of a farm boy so I got up about daybreak every day and saddled my horse and I'd ride out. Well, my horse loosened up a shoe so I rode back in to tighten... to get the equipment to tighten the shoe back up and get it going. Roger came in just a little bit later and he said, 'By golly, Bob being as your gonna go in a few days here, can we go out and stay all night up in the mountains in some of that area where it really looked like great habitat?'"

"I said, 'Well, yeah, I suppose.' So, we started out that afternoon, well, it was early afternoon... I suppose it was twelve... twelve-thirty... something like that. (We) took a little horse with us with our sleeping bags and our equipment all on it... rode up this creek bed about three- and one-half miles from where we were camped... Came around a bend by the crick where there was a ... I don't know how many folks here understand when I say downfall tree. There was a tree downfall with the roots system up. It was probably eight to ten, twelve feet high and that had caused the crick to reroute itself and go around there. I'm not real good on diagrams... Anyway, when we came around that bend in the creek, the creature stood across the crick about as far as far as that gentleman is right there with the paper in his hand (indicating a very close encounter). It was standing when I saw it. It may have been stooped down when Roger first saw it, but I was just a horse length or so behind Roger, leading the pack horse. When I saw the creature, it was standing up looking across the crick to where we were. But it immediately turned and started walking away... just taking natural steps. I mean natural looking steps to us which measure out forty-two to forty-eight inches from heel to toe... which is a pretty good step. There are some pretty tall men here and I'd almost dare them to make a step comfortably that long and keep on walking.

"That's how this film footage kinda got started. Roger's little horse kinda threw a little… I don't know what kind of dance you'd call it, but he wasn't… Roger was trying to bail off and get his camera at the same time… which he did. Roger was a very agile man… of course, he was a rodeo cowboy and he was a great athlete. So, he bailed off that little horse, with his camera and ran across the crick trying to get that camera focused on the creature. There was a little incline of sand alongside the crick on the other side. Roger kinda stumbled and fell when he got on the other side of that. He went down on his elbows. If you've seen that film footage from beginning to end, you'll see all that shaky part at the beginning. There is hardly… you can't hardly identify anything. Then Roger realized the creature was moving on and all this time I was setting right where I was. I stopped… the pack horse pulled loose or I let him go, I don't know. Everything happened so rapidly. I really don't remember if he pulled loose of if I just threw the rope back at him 'cause I was having a little hard time holding my horse. It was an older horse… it was an old roping horse and he was a little bit easier to handle than Roger's little horse that had a lot of spunk."

"There's when it all started going kinda wild. Roger wanted to relocate because she was movin'… I say she but I didn't know if it was a female or a male… it didn't make any difference, it was going, you know? Roger had to relocate to get a better… a better view, so he asked me if I would cover him. He told me later that he never… he was kinda concerned about two more being right close to this one. It was a wooded area on the far side… a really wooded area and you couldn't see up in there so that's what his concern was. Well, I didn't know that but he said 'Bob, will you cover me?' Well, I know I had a rifle in my scabbard on my saddle but I knew I couldn't do it sitting on a horse if anything happened that I needed to do because I was an avid hunter at that time and shot a lot of big game. I knew what I had to do, so I rode across the crick, stepped off the horse with my rifle in my hand… and that's where you see that famous turn when she turns and looks back at me… when I stepped down off that horse."

"Maybe the misconception was that I intended to shoot this creature… I did not intend to shoot the creature. I never raised the rifle to my shoulder… ever. I was carrying a 30'06… a Remington .30'06 with 180 grain bullets in it. I felt if I had to, I probably could stop this creature if it came back at us. As long as she kept walking away, I had no reason to even bring the rifle up. Like I said a few seconds ago, everything was happening so rapidly you just didn't have to do a lot. She was walking away all this time. Then, Roger hollered at me, he says 'Damn, Bob, I ran out of film…'"

"I got back up on the horse, I said, 'I'm going to follow…' and he said, 'no… no, no, no, no, no… Bob, don't do that, the other two might be here…' So, he was assuming this creature had… we were only about four miles from where those footprints (*Thom: these footprints mentioned here are the prints that drew the pair to Bluff Creek to begin with… there were three sizes of prints found there…*) were… Roger was assuming there would be two more there and he didn't want to be left there with just a camera in his hand… and no film in it. I mean now it seems kind of amusing to us but I don't think it was to Roger… when I seen the look on his face, he was glad I turned around and came back."

"By then, he got under a poncho. I didn't know much about cameras or anything… got underneath an old poncho that I had on the back of my saddle and got more film in his camera. Then we proceeded to catch up his horse which ran down the crick a ways and the little pack horse. And then we tried to follow where this creature went. We got up there a quite a little ways and we only saw one half of a wet print on a rock going across the crick and up the side of the mountain. It was steep and I kinda wanted to go after it, but I didn't know really why, I just wanted to see it again."

"Roger said, 'No, we gotta get back, Bob. Cause October twentieth your days are pretty short down there in the mountains.'"

"When that sun sets over them hills it don't take long 'fore it gets dark. By then, it was in the afternoon and we had to go back down to camp to get the material to make the casts. We did some things there."

"Roger had me get up on a

Bob Gimlin at OSS 2010

stump about, oh, three and a half or four feet high maybe… that was alongside of where she went and made her tracks. I had on a cowboy boot with a riding heel which is a pretty sharp heel now if you understand what that means. At that time, I was a little heavier than I am now… I weighed about one hundred and seventy-five pounds. I'd jump off beside this footprint… tracks… with that heel with one foot hit first and see how deep I could go into the soil that she walked through. It wouldn't go near as deep as her tracks were."

"Then I rode this horse alongside. He was a sixteen-hand quarter horse and he weighed about twelve hundred pounds. With my weight on him and the saddle I rode him right as close without disturbing the tracks as I could. Roger took pictures of that. The horse never made tracks as deep as the creature did. So that indicated right there that she was fairly heavy... that she was a heavy, heavy muscled creature which you can see in all this. "

"People say, 'What'd she weigh, Bob?'"

"I don't know I thought three or four hundred pounds was really big, you know? I said, 'probably three or four hundred pounds and, and six and a half foot tall.' Well, come to find out, I was way off on everything, you know? But, like I say, when I first saw her, I was up on a horse sixteen hands high which I was about nine feet up. Things will look a little different from that height."

"...You folks see what the result was from that film footage... that short film footage that we were lucky enough to get. At that point in time being as Roger fell down, we had no idea that we had any good film footage at all. Naturally we got the cast made and the pictures made of what we had to do there for what evidence we could get. Then we went in to mail that to Yakima or wherever he mailed it to. There has been a lot of controversy on where that film was processed and where it was mailed to. I never paid that much attention to it because I was very tired from being down there three weeks, riding horses every day long hours and driving the truck at night."

"Roger slept in a whole lot better than I did cause I just didn't sleep that much down there."

The Day of Destiny

What followed in the nearly half century to follow this momentous day has been an epic in incredulity. Expert after expert has testified to the veracity of the many aspects of this film. They have been largely ignored in favor of the statements of people who have no idea of what took place that day so long ago. Many people have even come forth to say that they were the subject in the suit… even though most could not identify where Bluff Creek even was, let alone how it could have been done.

Understand, there is no controversy among the experts who examined the evidence first hand at the time of discovery. The controversy comes from the statements of people who have no idea of which they speak and have no credentials to support their claims. Those who know the least are often those who speak the loudest.

With this in mind, this section is dedicated to the testimony of qualified experts. All statements contained are cited by name and credential. Now, to the facts…

Very early on in the Patterson-Gimlin Film, at Frame 5, an anomaly shows up in later copies on the right upper leg of the subject of the film.

Very evident in the frame, there is a bulge in the quadriceps muscle. This is not a gunshot wound as some have purported nor is it something someone would design into a suit being built for the purpose of a hoax.

Dr. Andrew Nelson of the Center for Motion Analysis and Biomechanics stated: "This is probably a rupture of the Quadriceps Muscle… this is something that cannot be copied in a suit."

He continued, *"After analyzing the biomechanical issues, I find it very hard to believe somebody in 1967 could have fabricated the intricacies as evidenced by the soft tissue irregularities seen on the upper leg. The science at that time was just far too primitive."*

John Chambers, now deceased, who won the Academy Award for costume design in 1969 for the 1968 award winning movie, "Planet of the Apes" stated: *"If this is a suit, it is the finest ever devised for it was beyond our capability in the 1960s. Every hair would have had to have been individually attached to the model for this to do what it does in that film."*

My question is very simple… Having heard what the experts in the field have said about this apparent rupture, does this sound like something that could have been accomplished in the remote wilds of northern California by two rodeo rough stock riders with no costuming or theatrical experience on a suit made of the hide of a red horse, as suggested by one fake Patty?

Fig 3 -3 Creature Maker John Chambers

Actually, that particular artifact in the film does not exist. It only appears in copies of copies of copies… some even to the eighth generations… Bill Munns, Hollywood creature maker and film analyst has stated that this does not EXIST in early generations of the film. He was privileged to examine and analyze a first-generation copy held by the widow of Roger Patterson, Patricia Patterson and that artifact simply is not there in that early generation! This is something that has become introduced by the copying and recopying of the film.

Size and Gait:

Professor Jeff Meldrum, Paleontologist, Idaho State University, stated that in primates the normal ratio of foot size to height is 6.5… that the foot is herein generally 15.5% of the height of the individual. It should be remembered that this is a "rule of thumb" only. There will be individuals who will not conform to this standard. That said, it should be noted that the general rule will still apply.

When Patty strode across that Bluff Creek sandbar, she left a very nicely defined track-way. Her prints were vivid and distinct, allowing the principles to cast the impressions using plaster of Paris as a medium.

Several noted authorities were called in to verify, independently, the scene. The measured size of the track as independently corroborated by John Green, Newspaperman from British Columbia, Canada, Bob Titmus from Redding, California and Al Hodgson of Willow Creek, California, all of whom saw and measured the tracks independently was fourteen- and one-half inches.

If we use the foot length as measured by independent sources, and apply it to Dr. Meldrum's formula we get a nominal height of nearly eight feet as follows:

14.5" Track X 6.5 = 94" = 7'10" in Height

Bill Munns, Graphic Artist, Hollywood set designer and analyst took a different approach to achieve a projected height for Patty. He had the Lens and Camera data and knowing the focal length of the lens, the magnification factor of that lens and the distance from the camera to the subject, he could ascertain the height of the subject of the film.

There was some confusion, initially, over what lens on the triple lens camera was actually used, but computations quickly solved that anomaly and Mr. Munns was able to state quite confidently that the being in the film was between seven and a half feet and eight feet tall as calculated.

The third method used to determine height utilizes the scene projected in Frame Seventy-Two of the film. We know that the length of the foot shown is fourteen- and one-half inches from the measurement of the tracks. If we allow for the shoulder slump, the bowed head and the knee bend of the right leg, and the fact that the right foot is sunken into the soil a finite distance, we can approximate her height by merely comparing the length of the foot as shown in the frame. I used a pair of dividers to apply the length of the foot to her over all height and came up with a height of seven feet two inches plus or minus three inches, or a ratio of about six point one times the foot length. That is certainly within the range to be expected from Dr. Meldrum's formula.

In this exercise we have used three different and distinct measuring techniques from three different experts to arrive at the same general conclusion: the subject in the film stood between seven feet four inches and seven feet ten inches. In no case did this study yield a figure that could be construed as the height of a normal human being found in society in 1967.

Gait:

The next area of study this treatise on height leads us is to the unique gait exhibited by Patty in the film. In order to shed some light on this area of interest, we must borrow from the world of make believe again and marry that to the world of science. Reuben Steindorff, Senior Animator at "Vision Realm, Inc.", using a system known as "Inverse Kinematics and Motion Analysis" created a 3D model of the subject in the Patterson-Gimlin film. This model was then forwarded to Dr. Andrew Nelson of the Center for Motion Dynamics and Biomechanics who graphically inserted a skeleton into the model of the creature.

Dr. Nelson ran an entire spectrum of tests on the test subject and determined that Patty walked with a "Compliant Gait". We humans, on the other hand, use a very stiff legged gait in which we, essentially, pole vault over our leg and land with a very heavy heel strike. Motion Capture analysis shows that the Compliant Gait results in a very smooth, swinging stride with virtually no heel strike and very flat-footed placement of the foot on the down step.

Dr. Scott Lind and Emmy Award winning animator Joe Russo then attempted to train an athlete to walk with this same compliant gait and found that the human body was not capable of exactly duplicating this movement. They were unsuccessful in their attempts to train even an athlete to walk with this gait.

This failure caused these professional men to conclude that it was impossible for a human to exactly duplicate the walking motion of the being in the Patterson-Gimlin Film.

Costuming:

An effort was made to analyze the possibility of the use of a costume in this short film. To this end, Peter Brooke, Costume Designer for the "Jim Henson Creature Shop" and famed Hollywood Costumer, John Chambers whose efforts on the "Planet of the Apes" took four professional designers three months to create, performed this examination and a thorough analysis of the being in the Patterson-Gimlin Film.

These consummate professionals concluded that there are three notable features in the film that needed to be closely examined and on which the conclusion depended. These three factors are:

1.   Arm Length
2.   Firm Musculature beneath the surface
3.   The Hair Adheres to the body beneath it

We shall examine each of these points individually as we continue in our quest for the real facts as pertains to this film. It seems that all who I have ever heard denouncing this as a person in a suit do not know the facts that these experts have presented in their analysis.

Bill Munns Graphic Artist

Peter Brooke, after his examinations, stated unequivocally, ***"Such Costumes did not exist in the 1960s." The fur adheres to the form and contours of the body. Today we make such suits of four way stretch fur fabrics but that did not exist until the 1980s. The era of that film did not have fur that could be form fitted."***

John Chambers added to Mr. Brooke's conclusions, ***"It does stretch. I don't know how they could have done that in 1967. There are several individual muscle groups that are plainly visible on the creature in the film."*** There is *even "tightening and slacking of the Achilles Tendon evident" as she walks.*

Mr. Chambers continued, stating that the *"Shoulder blade is clearly visible and moves during the walk and the look back."*

I believe that if I stopped here and ventured no further there could be few who could argue successfully that this film could in any way be contrived or faked. All of the factors mentioned to this point are patently visible and are easily seen by even the most ardent of skeptics… whether those skeptics would admit to seeing these factors is entirely another question and for that reason, if no other, we will go on with our treatise.

Bill Munns, Graphic Artist and Designer most noted, probably, for his large sized artistic representation of the ancient hominid, Gigantopithecus blacki.

Mister Munns has completed a thorough analysis of the requirements of a person wearing a suit such as one shown in the Patterson-Gimlin film would have to be. He has also completed an involved diagnosis of that film. I would recommend that anyone interested in this subject read his book, "When Roger Met Patty", visit his website at www.themunnsreport.com and his series of Youtube videos at:

http://www.youtube.com/watch?feature=player_emb edded&v=cJZTLWUJh-w#!

This is the first of a series of videos he has created concerning this subject. A treatise on the comparative anatomical features of a female model and the image of Patty is shown clearly in the video at:

http://www.youtube.com/watch?v=V9WO8c38cRo&fe
ature=relmfu

In this work, Bill Munns carefully illustrates the motions of the person doing the filming and the subject of the film as they move through the setting of the encounter. He carefully illustrates the relative positions of both entities during this time frame. Part Two illustrates the track-way and the individual tracks therein. Part Three is the most important for our purposes. In this segment, Mr. Munns analyzes the subject in the film, Patty, and compares that to an actual female subject.

In Bill Munns' characterization, he points out numerous anomalies. He illustrates with side-by-side comparisons the impossibility of creating any suit like that which would have necessarily have to have been used in making the Patterson-Gimlin Film. His comparisons illustrate many incompatibilities inherent with placing a human into the anatomy of the subject of the film. His statement is that the anatomical differences leave us *"with an impossible costume to construct..."*

Mr. Munns stated, "The subject in the film has anatomical proportions that are very odd to say the least. Conventional creative costume design and tricks or illusions for altering regular human proportions do not achieve the result easily or effectively. Now, there are many ignorant people who have no knowledge of designing or fabricating creature costumes that will try to tell you differently because, in their ignorance, they fail to understand the difference between what is theoretically possible and what factually practical to accomplish. The simple reality is that... the human female shown could not be dressed up in any fur costume and make a perfect functioning performance equal to the Patterson-Gimlin subject performance. *The challenge of putting a human into a fur costume to fake the PGF is far more difficult and questionable than the hoax believers claim that it is."*

In short, Mr. Munns stated that "You cannot alter where the knees or elbows bend. She has long upper leg and short lower leg." Due to the structure of the creature in the film, *"If you could find a suit to match all the criteria necessary (a suit other experts have testified was not possible in 1967), <u>you could not find a human who could wear that suit."</u>*

Arm Length Vs. Leg Length

Property of BFRO – Used With Permission

Dr. Jeff Meldrum, Paleontologist, Idaho State University has taught a ratio known as the Intermembral Index or IM. Basically, the IM is the ratio of the arm length to the leg length of a subject times one hundred (to remove the decimal). In Humans this IM ratio is 72.

In 1998, the British Broadcasting Company, BBC,

90

approved and financed a program for their network to "disprove" the Patterson Gimlin Film by creating a state-of-the-art costume and putting a man inside to prove that it could be done without a real sasquatch having to have done it. It is true that this project was being done over thirty years after the original film came into being. This new film would be taking advantage of all the latest technology available, with no regard as to its availability in 1967. They theorized that the people who made us believe in Wookies and Ewoks, Apes that ruled planets and all such creations would surely debunk the claim that the suit used in that long ago film was impossible to duplicate in these conditions... This figure is that suit…

There are two major differences in the suit pictured on the right with the man inside and the picture of Patty that are readily evident. First, pay close attention to the head and shoulder positions in the turn to look back.

Patty's head sits very low on her shoulders. Her neck is very short so that when she turns, her chin goes into her shoulder, preventing it from rotating further to the rear. The result is that the shoulder has to be rotated back and out of the way for her to swivel her head far enough to effect the "look back" seen in frame 354 of the film (erroneously reported as frame 352 due to inaccuracies in counting the frames). In the right suited figure, this is not a problem as the head sits high atop the pedestal that is the neck and can swivel over the top of the shoulder as it is shown doing here. The difference is quite obvious. The second obvious anomaly is created by the difference in the length of the arms. In that same right suited figure, the right arm is in approximately the same position as its swing ends above the hip. That individual's hand is above the level of his buttocks. Compare that to the length and position of Patty's arm and hand. Her arms end well below her buttocks, not above as in the right suited figure. Once again, as in all such endeavors, not only was the Patterson-Gimlin film not disproved, but, in essence, it served to offer very good evidence for the veracity of that film. Aside from the fact that the person in suit could not begin to duplicate the compliant gait of the actual Sasquatch, a mere glance shows a very glaring error. Please note the relative arm lengths in the two figures. The right suited fake has arms that, in relation to the size of its body are HUMAN in form. The black figure, Patty, from the film frame 354, has arms that are much longer in relationship to her body. In fact, BBC, after airing this show issued a disclaimer stating that the data contained therein did not refute the claims of the Patterson-Gimlin Film.

Intermembral Index

Dr. Jeff Meldrum describes the Intermembral Index (IMI) as the ratio of the arm as measured from the shoulder to the wrist, to the leg, measured from the hip to the ankle times one hundred. The one hundred factor is simply to clear the decimal from the result. Mathematically, that is Arm Length divided by Leg Length times one hundred or:

$$AL / LL \times 100 = IMI$$

In primate species, all members have a distinct and specific IMI. They break down as follows:

In a human, the IM is 72
In a chimpanzee, the IM is 108
In a gorilla, the IM is 122

In a sasquatch, the IM is 84

As can be readily seen here, the IM of human and sasquatch are remarkably similar but certainly not identical. It is this difference that can be used to make qualitative evaluations on reported sasquatch photographs and videos.

If we return to frame 72 of the Patterson- Gimlin Film again, the arm length and the leg length are readily apparent and can be easily measured. Also, the arm and the leg are at the same distance from the lens in the same plane so no distortion or foreshortening is introduced into the exercise. The previously discussed calculation using the formula provided by Dr. Meldrum yields an IM of this figure to be 84. That measurement and calculation places it firmly into the range expected from its species.

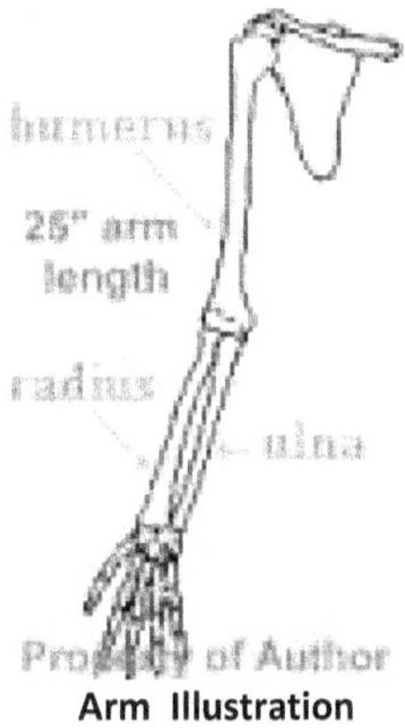

Arm Illustration

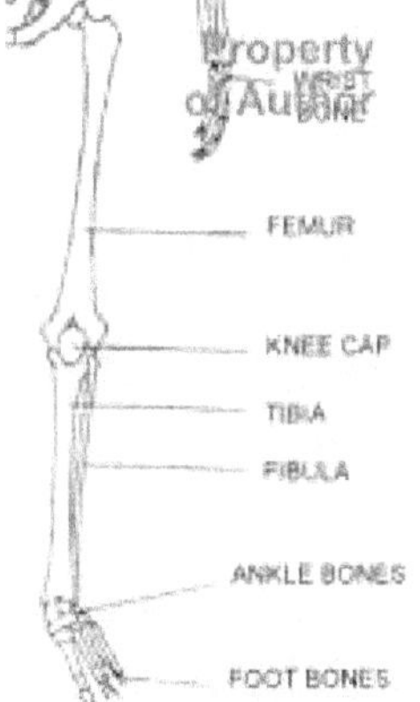

Leg Illustration

At this point the ardent skeptic, the sort that this treatise is designed for, would state something like, "Oh they just used an arm extender to make it correct…" Let us examine this more closely.

I measured my arm and came up with the following data:

Arm Length = 25"…
Leg Length = 34.5"…
therefore, my IM is: 25/34.5X100=72.4…

While this places me most firmly as human in structure, it creates a problem for he who would use an arm extender to achieve an IM of 84 as evidenced by the figure in the film. Solving the equation to yield a known IM is as follows:

$$(AL / 34.5) \times 100 = 84 \text{ or, } AL = 84 \times 34.5 / 100 = 29$$

This means, simply stated, that my arm length of twenty-five inches would have to be extended by four inches to achieve the desired result. In viewing the figure here, where would those four inches be added?   If it were done as in the figure above, that would make the lower arm well out of proportion to the upper arm. In Frame 72, it is very evident that the elbow is exactly where it

should be placed in the arm. As has been stated prior: "The elbow cannot be relocated."

It simply is where it is.

A common ploy to trick the eye is to use oversized hands in the form of huge gloves to convey the idea of a longer arm. There was a recent video released by a fellow attempting to "disprove" the film… that film used this ploy, but the result was so emphatically terrible, it is beyond reason to ever believe it could be even remotely possible. Look at the hands in the film and it is readily apparent that they are in proportion to the rest of the body, not some out-sized grotesqueness perpetrated by one who would have us believe his tripe.

There is one other method for achieving the desired ratio. Perhaps one could reduce my leg length to the requisite twenty-nine inches by somehow removing five and a half inches from its present length. Somehow, I think that might be even more objectionable to me than the process of adding the needed length to my arms. In short, there is really no feasible way short of surgery to alter this Intermembral Index formula or measuring criteria.

Some Comparisons

On Bill Munns' Youtube channel, he has a complete workup on the dynamics of the comparative anatomy of the Patterson-Gimlin entity with a human female. While Bill goes into great detail in that video series to prove the individual in that film cannot be a human in a suit, I will cover only one aspect of it here. Anyone who wants to see more can simply go to www.Youtube.com and do a search for "The Munns Report 3A" and view it at his leisure.
What this illustration shows is a female human in comparison with Patty, the individual from the Patterson-Gimlin Film, scaled to the same height and the lady's right leg and left hand adjusted slightly to approximate closely those of Patty from the film.

Two reference lines are drawn for comparison of the two figures, one at the shoulder lines and the other at the tip of a closed right hand. As can be seen, the two lines correspond on the two individuals. Notice that on the human female, the hand line passes just below the dot indicating her hip, yet is well below the buttocks on Patty. It must needs be noted here that putting a furred costume on the human would exacerbate the problem as costumes ENLARGE the figure, not reduce it as would need to be done here to make the suit fit.

The next line of interest for us is that directly above the line at the shoulder level… that which passes at Patty's brow level and passes below the nose level of the woman.

The next line of interest is the uppermost line that is at the top of the woman's head, but passes over the top of Patty's head. Further, if one looks the inset of the profile of the

head, it is obvious that, while the woman's head goes

straight vertical from her brow, in Patty, that forehead slopes back so significantly that the woman's head could never even fit into a mask made to the proportions of Patty.

Finally, we reach a logical conclusion… Even though too many people grasp onto the "theoretically possible" as something that practically doable and try to equate them, such is not the case and, in this particular case, even if we had materials not available for a decade later to make this suit, we could not find a person that could fit into it! Clearly, Patty cannot be a person in a fur suit!

Test combo
(See chapter 1 for the purpose of this picture.)

Just past Mile Post Fourteen it happened... and we witnessed a most interesting event. There were three of us in my car. I was driving, Jackie, having come here for the Ronnyvoo from far-off Devon, England was riding shotgun in the front passenger's seat with Arla directly behind her in the REO position. At the time of this event, Arla had her eyes closed and witnessed only the aftermath. Jackie was alert and I was extremely alert, given the driving I was doing.

At a fairly good rate of speed, we rounded a curve and transitioned from sunlight to full shade. It was nearly seven pm so the sun was at a low angle in the sky and our west to west-northwest route had been giving me fits with the sun in my eyes, making the transition to shade most welcome!

As soon as the transition was complete and my eyes adjusted to the new light level, I saw two figures standing directly before my car and a bit over half way to the right-hand side of the two-lane roadway when viewed from my vantage point. This placed them well into the right quadrant of the roadway. The two individuals immediately turned and, with but two paces, crossed the roadway back to my left and exited it on my left side.

My initial thought was "BEAR", but that was immediately dismissed as they were both erect and moving bi-pedally. My next thought was "MAN"! That died aborning as these were much too large to be men. I was looking UP at them from the seat of my car at a range of less than twenty yards and more probably at a range that would not exceed ten to fifteen yards. It should be understood that these thought shifts were immediate! There was no consideration nor deliberation involved. My mind merely cycled through these options, discarding the inappropriate and storing a living image of the conclusion reached. The entire cycling, I'm sure, did not last a full second in real time.

Instantly, I knew what I was seeing! They were large… very large... exceeding seven and a half feet and probably attaining more than eight feet in height. They were entirely hirsute, except for their faces which were remarkably bare, and their feet, they were completely covered in dark, steel gray and black pelage. Their shoulders were humped as they moved rapidly and their Ostman Pads shone brightly as they made haste to flee from our sight. It was the work of but moments for them to cover the width of that road with the second step taking them over the side of a bank that exceeded one-hundred-forty percent slope. In the blink of an eye, they were gone, but for that instant their image burned a scene onto the retina of my eye. That image has not dimmed with time, but is as bright as it was that day.

For just a few seconds we sat and reflected on what we had seen... the sound of Jackie's exclamation of "Bigfoot" was still ringing in my ears when a most amazing sequence of events was set into motion. Jackie, from her shotgun position, saw them as quickly as I did...

Fig 4-8 Turtle Rock

She recognized them immediately for what they were...

Kathi's Call

By this evening, number seven in camp, we were nearly immune to the foibles and follies of the indigenous hairy people. Jackie had retired and Kathi and I remained by the fire awhile longer in hopes of more intimate contact. Finally, the long day began to exact its revenge on us... especially me... so I bid her a good evening and retired to my tent and my ritual of preparing my evening medications, my bed and my reading material.

It was the work of several minutes to accomplish all necessary to allow me to retire. My medications were taken, Insulin injections complete, clock wound (I hate awaking in the dark of night and wondering just what the time was...), heater lighted and clothes and paraphernalia arranged and stored against the night. Since, on this night, there were to be but two of us in my large tent, this storage was a bit more haphazard than is normally the case.

There were two items that caused me to smile when I arranged my tent for sleeping. First is the beautiful Father's Day card I received from my newest friends, Nancy and Russ "Gizmo" Cobb (I have never seen ANYONE so adept at making something utile and desirable from next to nothing in my LIFE!). The other is a stone... just a piece of igneous rock with an inordinately high iron content that called to me as I moved through camp early on in our stay. It should be noted here that stones and I have long felt an affinity for one another to the point that, when younger... in the middle years of the last century... it was not uncommon for me, after a particularly exceptional day of feeling "called" by one stone or another, for my jacket pockets to arrive back at my home a full ten to fifteen minutes after I did! Now that said, I smiled as I moved this "special" stone to its night resting position from its daytime post.

My nest had been built and I was just undressed but not yet in my bed when it came out of the night... it was loud... as loud as any utterance I had heard this trip, causing me to immediately climb back onto my feet...

"THOM... THOM..." the cry resounded off the surrounding cedars... "Help me Thom... I can't move... Oh my brother... my sister... I'm so sorry but you scared me! I didn't expect you to be there... THOM! I need help now! Oh, my dear brother or sister, I'm so sorry... Thom, where are you?"

"I'm coming Kathi," I called out to her... "Just let me get my pants on and I will be there in a moment..."

"Hurry... I can't move and he's HUGE... Please hurry!"

For all those who know me, it's understood that my "hurry" mode is not all that fast with the exigencies of age plying their wares on my physicality, but I really did do my best to speed my way

Fig 4-10 Kathi's Tent

along... even to the point that I disregarded the need for such amenities as shoes, socks or jacket in my desire to rescue this damsel from her durance vile.

It must be understood that, until now, I had no idea of what this threat was constituted. I knew not what the plague consisted of nor did I know how close disaster actually was. My first thought was "bear" as we had been seeing so many of them in our travels. Nancy had even had an encounter just up the road from camp no more than a hundred and fifty yards from Kathi's obviously now beleaguered tent. Her appealing to it as "brother or sister" tended to make me disbelieve this bear theory, as I knew of no such beings in her close family.

Kathi is not one to panic at shadows on the wind. She has had encounters with the sasquatch people before and had met them logically and forthrightly. If she was now exorcised, there was reason, so I kept up a constant conversation, albeit one way, with her as I made my way to her side. I'm not sure she ever heard me as she continued her own stream of apologies to whatever was causing her distress and not 

paying me a whole lot of attention. Presently, I arrived at her side... not more than fifteen feet from her tent and possibly as close as twelve feet. As I watched her while approaching rapidly (for me...) I raised my eyes to follow the direction of her gaze... and saw...

"Oh my gosh," I uttered in a barely audible tone. Then, more loudly, "Well, hello, Big Guy... How are you tonight?"

As Kathi had left the fire pit after having

disassembled the fire and walked the few yards to her tent, she was well at peace within herself. As she passed the few small trees that grew beside the road at the edge of the green meadow where her small tent was installed, she flashed her light in the direction of her tent and audibly though quietly said, "There's my tent..." as the light flashed across the reflective material on the front of the nylon. There was no response of any kind, nor was any such required or expected. When she had taken but a very few more steps toward the structure, she again raised her hand light to illuminate it more fully.

What occurred then was most remarkable. As the light shone forward, a large, dark body reared up from just before the shelter. They were eyeball to eyeball at not more than ten feet separation. From my position inside my tent in the pursuit of making up my bed, I heard his grunt but did not, at that time, register it as anything more than one more greeting in a plethora of such I had so far absorbed this week.

The large fellow rose to his full height... probably as alarmed by the encounter as was Kathi. Our girl, however, could not raise her light at all. She was locked on to his waist area and could neither elevate her light nor her eyes any higher. As such, she had a very good and solid view of him from the waist down only. She had seen his torso as he stood, but could no longer focus on it.

Kathi was frozen in place. She could move

Kathi

no part of her body whatsoever. When I arrived at her side and greeted him, she was locked in her stance. Her head was low and her eyes cast downward. Had I not have greeted him verbally, she would not have known I had moved up behind her.

...But I had... and what I saw in my light, dim though it was, was a large, very male sasquatch standing beside a twenty foot, or so, tall cedar tree watching us intently. Kathi continued her exhortations, apologizing profusely and sincerely for having startled him. At the same time, he was attempting to apologize to her and the ensuing non-communicative conversation was quite amusing to me when seen from the outside.

Eventually, she calmed as did the big fellow and by taking her arm and telling him that I would assume responsibility for her, he released his lock on her and I instructed her to get her kak from the tent and move it into mine for the night. It is important to understand that this physical lock he used is a positive thing. If she were allowed to run at will, she might well have fallen and injured herself badly. Perhaps she would have panicked and run directly at him or into the path of others in the vicinity. By locking her up, he was able to render her safely unable to move and to make his exit without further harm or trauma.

Although the clan remained with us all night long, poking and prodding our tent at times, no further incidents of this magnitude occurred and before morning, Kathi was up and doing even if it was a bit slower and more studied than normal!

From Kathi's Viewpoint

We have a unique opportunity with this incident because we have it written from two different views… the views of the two humans who witnessed the events. I have given my account of that mid-June night at Elk Meadows, at the end of the eastern G-O road in 2014. Now, we shall hear the same event described by she who lived it. Each account was written separately and each appeared in books authored by the respective individuals. Mine appeared in "21 Days to Destiny – The Real Story of Bluff Creek", while hers appeared in her book, "My Journey into Myth and Mystery".

Kathi's account begins with a description of events in the night that occurred almost nightly in this spot. It should be made clear here that Kathi's tent was set up a bit remote from others in the group as she solicited the night activities around her tent. That said, her tent was about fifty yards from mine, directly measured, but it sat back in a little alcove of the meadow, with a great stand of Port Orford Cedar trees behind her.

Her Account

That night I got into my tent and was reading with the light on when I heard the sound of footprints approach my tent. I quit breathing and sat still to see what they would do. They had to know I was still up because my tent was lit up like the mothership at the time. The footsteps stopped a few feet from my tent and then I started to hear the sound of vegetation being plucked. Not ripped up out of the ground, but plucked. Like when one sits on the ground and plucks out blades of grass, one at a time

106

This really confused me. For a brief time, my mind tried to tell me that it was the deer I met here earlier and she had come back to eat some grass. But I knew beyond a shadow of a doubt that I had heard the sound of heavy footsteps and nothing else in the world sounds like that. Especially, when they are so heavy you can almost feel them on the ground. So, I laid there trying to figure out exactly what was going on beside my tent… though I will admit that I never once contemplated unzipping the tent and poking my head out for confirmation of what I was hearing.

This went on for quite some time and I started getting tired and wanted to climb into my bed, but once again, I didn't want to make a wrong move in case this somebody had thought I was asleep. I'm pretty sure it's bad form to surprise or irritate a species this large! The longer I sat and listened, the surer I became that it was the sound of vegetation being plucked. Then, whatever was out there, walked back into the tree line and I never heard anything afterwards

The next evening, I heard the sound of heavy football around my tent. This time they were coming in from both sides of the tent and I felt a bit intimidated by them. They were heavier than what I had heard previously and the thought that went through my mind was that the adults had just entered camp. I then heard a "tink" sound… like a rock hitting a rock. It was a small sound and it came from behind and to the left of my tent. As soon as I heard it, I knew exactly where it came from. Under the tree directly behind my tent I had spotted an area with rocks laying in the dirt. It had caught my eye because the rocks were chipped in various spots and had rock dust laying on a few of them and it looked like they had been hit together in the past.

A few seconds later I heard it again… Tink. I just smiled and waited to see if it would come again. I did it… Tink. Each sound was exactly the same and came from the same spot. I remember wondering what they could be doing back there. Then I heard footsteps coming from in front of my tent, continuing on the longer side of it, and then heading back towards where the tink sounds were coming from. I heard rustling sounds from the tree back there for a while, but never heard any vocals of any kind. I then heard a branch snap come from way back in the tree line. At that sound, both sets of footsteps walked away from the back of my tent, went up to dry creek bed and off into the trees.

One night, before putting the fire out, I stood in front of it thinking about how I was in for a long night. The sasquatch had been spotted on the hill across from our camp kitchen during the day. I knew that if they were getting brave enough to be seen during the day, they were definitely not afraid to come on into camp at night and do whatever they wanted to do. I had no regrets about pitching my tent so far away, but it sure was a long walk to get there after I put the fire out each evening.

I didn't light any lights when I got into my tent that night. I was pretty tired and the tent was lit up enough that I could actually see well enough to get ready for bed and climb in my bag. While I was laying there, I once again heard the sound of something large walk up to my tent, and this time they came all the way up to the front corner of it.   I then heard what sounded to me like branches rubbing across the top of the tent, or perhaps pine needles scattering across the top of it. I looked up to see a shadow in the shape of a very large hand and arm making a slow circle across the top of the tent!

Gridlock

A few of our fellow campers decided to stay in Willow Creek, but three others but three of the other group wanted to come up to our camp for a night to get in on the action we were having there. They decided to drive up the following morning, so we left and made our way back to work camping area.

We didn't get back to our camp until nearly sunset and it was so nice to pull in and see a fire already lit at the camp kitchen for us. Russ and Nancy had come back earlier than we had and had taken the time to get it already for us and we were ever so grateful for that. There were only five of us in camp now that the others had stayed in Willow Creek and we sat around the fire talking.

Eventually, everyone wandered off to bed and Thom and I sat at the fire listening to the sound of screams and wailing coming from down the mountain behind us. It was the first time I had heard that type of vocalization and I could certainly see how scary that sound could be if you were alone in the forest. They were far enough away that I could just get the sound of it, so I assumed that they had business elsewhere and we would be alone in camp that night. I didn't stay out by the fire for long after Thom said good night as I was pretty tired myself.

There was no moon present in the sky that night and it became very dark after the sun went down. Instead of just cutting across the grass to get to my tent like I had previously, I elected to walk out into the dirt road that cut through the middle of the meadow and follow that to my area. I didn't want to chance turn in an ankle tripping over the many rocks that jutted out of the grass.

There are trees growing along one side of this road which ended abruptly at a dry creek makes a bend and comes down off the mountain. It is very deep with large rocks at the bottom of it. It's impossible to see into the side of the meadow where my tent was from the road until the trees end and you were instantly dumped into it.

I had turned on my flashlight as I was walking down the dirt road, but kept it hanging down towards the ground to light up my feet add the road in front of me. It wasn't my wand light, which would light up a football field, it was a mag light with only a small circle of light. I hadn't heard any sounds at all since Thom and I had heard the wailing earlier in the evening. There were no footfalls nor whistles, so I assume that we were all alone out there as far as the Sasquatch were concerned. Just goes to show how quiet and elusive the species can be when they want to be! I'm still not sure how they can be so loud and you can actually feel their footsteps hitting the ground when they want you to, and yet be so quiet and stealthy that you never even know they're there when they don't want you to.

I walked out of camp with my flashlight pointing down at the ground and walked up the dirt road along the tree line. At one point I remember stopping and just standing there and looking up at the stars. They appeared to be so close you could just reach up and touch them. I remember saying how beautiful it was up there, and how much I was going to miss it when I went home in a few days. I didn't want to think about leaving just then, so I dismissed that thought and continued walking up the dirt road to my part of the meadow.

When I arrived at the little tree, I lifted my mag light and shone it cross the field. I moved the lights slowly from one end of this area did the other end twice... Once at ground level and then back over it again, a bit higher up. I was actually looking for eye shine from the bear or a deer as I wasn't expecting any Sasquatch to be out there. The first time I swept the meadow, I saw my tent reflecting the light back at me and there was nothing in front of it or near it. There was nothing on the ground in the meadow. The second time I swept the meadow there was my tent reflecting back at me with nothing at all in front of it, or near it at approximately five to six feet off the ground. I didn't turn the light up any higher or up into the tree beyond my tent.

As my light hit my tent the second time, I remember saying out loud, "Oh, there's my tent," as I wanted to talk out loud in case there was a bear in the tree line. I knew in my heart that there was nothing out in the middle of the meadow because I was able to see every flower and every weed growing on the ground there. I was taking a close look for a bear or deer or any ocean that I might see before I walked out there, and I thought I had scrutinized the area really well.

I was still standing on the dirt road at this point, and hadn't completely walked out into the open area beyond the last little tree. That would be the new growth that seen to the right of the in the picture. Once I was convinced that there was nothing eating out in the meadow, I walked past the two new growth trees and was now standing to the left of them directly in front of my tent. (That would be where the gray rock pops up out of the ground to the left in the picture)

I stopped at the top of the runoff trench to quickly scan the bottom with my light so that I could take stock on where all the large boulders were located, and then continued sweeping my light up the other side of the trench to calculate exactly what path I was going to take to get through here, all in one continuous motion. It wasn't a wide trench so this took perhaps a fraction of a second to maneuver. At this point, I still had not heard a single sound out of the night. It would've taken me approximately eight steps to get from there to my tent.

As soon as my light came up out of the trench and hit the front of my tent, I saw a sasquatch standing there. In the wink of an eye, it turned, and sort of hopped over the corner of my tent and ran towards the tree that sat behind it. All I really remember seeing were two large trunk tree trunks turn (or swivel) and one tree went around the left corner of my tent and the other one followed it. They then they took three steps and stopped behind the tree. I not only saw the steps, but I felt them hit the ground. It's hard to actually describe such a heavy, flat, thudding sound to people, but I will never forget exactly what it sounded like.

I don't remember looking into his face, but I know I did. I just can't bring it to the front of my mind at all. The only thing my mind wants to remember seeing is its legs moving around the corner of the tent. My mind wanted to identify that as tree trunk because the legs were easily the size of a medium tree, and about the color redwood. But when my eyes caught sight of the ankles… And we all know the trees don't have ankles! I then saw the hair at the bottom of the calf float up slightly as it raised its leg and put it back down again. It was the most awesome site I have seen to date. This baby was moving swiftly too!

I say that I saw two tree trunks because that is exactly what my mind wanted to describe them as… thick, redwood color, tree trunks. Except for the fact that these tree trunks had ankles! The legs tapered down and I saw definitely saw ankles is it lifted its leg to hop over the side of the tent. And these tree trunks were covered in hair that swept up above the ankle as it ran away… Not like it was blowing in the wind, but just slightly puffed up away from the leg as it turned and jumped over the edge of the tent. I think that the reason I remember this part so vividly was because my mind worked so hard try and identify them as tree trunks, and it was having a really hard time doing so. I would love to be hypnotized and see if I can remember what face looked like.

It all happened so fast that I barely had time to log it all, but the one thing that my mind did grab hold of was the fact that it had only taken three steps and then add planted itself right behind that tree. And that tree was only sixteen steps away from me. I know this because my tent is eight by eight and the tree sat directly behind it! I stood and listened for it to continue running off in the opposite direction, but he stayed right there watching me.

I know I must have stood stunned for a few minutes just gathering my thoughts and waiting for it to run off. When I finally did come to back to my senses, my first thought was that if it wasn't leaving, then I better apologize! And that is exactly what I did.

I said, "Oh, I am so sorry brother or sister! I didn't see you standing there, or I would never have shown the light at you. I'm sorry to scare you, but you know you just scared the heck out of me too! I think I just lost about a year or so of my life, but I am so sorry if I scared you! If you want to stay here in the meadow and eat, or whatever you were doing, it's just fine by me. I'll just go back over to the fire and let you do what you came here to do! I am so sorry if I scared you, but you have to admit that we just scared each other."

I chuckled them to let it know I was still in a good mood, and maybe that would loosen any anger he/she may have been feeling.

At that point, I had intended to just swivel around and head back to the camp kitchen area to build up the fire again and let them have the meadow. I had absolutely no problem with that thought at all but, as I went to swivel around, I noticed that I couldn't do it. I tried to get my feet to back me up so I could get out of there, but they wouldn't move either. I started to get a bit frustrated at that point and I again started blabbering out loud and said "I really am sorry brother or sister and I didn't mean to come up on you like that. I want goodwill and I mean you no harm. I'm going to go back over to the fire and leave you alone in the meadow now. "

Once again, I found that I could not get my body to move on command and I noticed that my head was now looking at the ground and my hand was pointing the light down toward the ground also. I have to admit, I was starting to get quite irritated at this point, but even I am not hardheaded enough to snap at something that is as potentially dangerous, so I did the only other thing I could think of, and that was the call Thom.

I yelled out, "Thom! Could you come here please? I need some help out here! Right now!" I knew that he was already in bed and I was quite happy to hear his voice ring out in the night telling me that he was on his way.

I was still apologizing profusely, and stuck in place when Thom arrived. When he came up behind me, I heard him say something, but all I caught was something about how big this one was. Well. I knew that, but I sure appreciated that someone else was able to see it and substantiate my claim. I just hoped that Thom could talk to this one and see if he was okay, or if he met us any harm. Thom began talking to it and I asked him if he was male or female, and he replied that it was a male.
I then started saying, "I am so sorry brother, it wasn't fair of me to walk up on you like that. Don't be mad at me or take it out on me, okay? I'm Sorry that we scared each other."

Yes, I was babbling at this point, and no, I didn't notice how much at the time. It just felt good to be able to address it specifically instead of saying brother/sister all the time, and I wanted to make it more personal for him. At one point Thom began to chuckle and I asked him what was going on? He replied, "You're both apologizing to each other at the same time and neither one of you is listening to the other and there is a complete lack of communication here."

That made me stop and think what was going on, and I sure wish that I had learned to open my mind up more so that I could communicate with this guy. It made me sad that I couldn't hear him. Thom told him that I was going to walk over to my tent and get my stuff and I would sleep with him in his tent that night. He gave me a slight nudge while leaning in towards me and whispered for me to walk over and get my stuff from my tent. I rocked forward a bit when he noticed me, and replied that the reason I called him was that I couldn't will my body to move for me. I was stuck!

He then spoke to the sasquatch again and told him that he would take full responsibility for me I would he please release his grip on me so that I could get my stuff out of the tent. It was the strangest thing that has ever happened to me when I felt my body go limp. I had to correct myself and take hold of my body again so I didn't just drop to the ground. I was bulging with electricity. My entire body was tingling like I just been shocked.

I handed Thom my walking stick and my flashlight and walked to my tent. I unzipped it and knelt in front of the door. I just pulled my bag from the tent and zipped it back up. I then walked back to Thom and we continued back across the meadow to his chair. The sasquatch never moved the entire time and I don't remember ever hearing them make a sound either.

What I got myself situated in Thom's tent and was lying there thinking about what it just happened, I heard Thom ask me if I was okay. I told him that my body was still tingling like I had just received electric shock therapy, but other than that I felt fine. He then said out loud, "Yes, she is doing okay now."

I asked him what that was all about and he told me that the sasquatch was checking in and asking if I was okay. I thought that was about the sweetest thing ever! To actually care about my feelings! I know humans who don't care that much about me, and it touched me deeply.

I asked Thom if he knew his name, and he told me what it was. I sure wish then that I could speak to him and tell him how much it had touched me that he was worried about my well-being, but all I could think of was his well-being and had been hoping that I had not offended them in any way. My fingers and my toes were the only things tingling on me at that point and I had to keep rubbing them, trying to make the feeling go away.

After Thom fell asleep, I could hear numerous footballs around the tent and out around the camp kitchen area. I then heard the hummingbird/bee sound moving around the inside of Thom's tent. This time I got a smile on my face and I said very quietly, "Oh sweet! That sound was from you guys, huh? Thanks for checking up on me, I'm doing fine. I hope you are too!" I then heard a brief communication happening outside and I listened to them converse for a few minutes.

Fortunately, Thom's tent is the size of a small cabin, so there was plenty of room for him and me along with Jackie to sleep quite comfortably together. Unfortunately, I had forgotten to grab the pad beneath my bag, so I felt every rock and every stick on the ground underneath me that night and I never really did get any sleep. As soon as I saw the sun light up in the sky, I grab my bag and headed for my tent, so that I could get a few hours of sleep before anyone else woke up.

All too soon I heard voices approach my tent and stand next to it talking about what had happened the night before. As soon as I opened my eyes, I knew it wasn't going to be my best day ever. I think I got about three hours sleep total and awoke very blurry eyed. Just trying to get my shoes on was the hardest work I had ever done. I'm not proud of how I must've looked, falling out of my tent as everyone was standing there looking at me and then started asking me questions...

Chapter 5
Witnessed Events
My Bigfoot Encounter
By Todd M. Neiss

Few moments in life have such a dramatic impact on a person's life that it qualifies as an 'epiphany'; a moment where one's concept of reality itself is utterly and permanently altered. Such a moment happened to me one sunny day in early spring of 1993. April 3rd was a day which is, and will always be, seared in my mind as if it were yesterday. Even after more than 18 years, the specifics of that event should illustrate the impact it had on my life.

It should be noted that, with regard to even the possibility of these creatures existing, I was beyond skeptical. Simply put, I had relegated these beasts to the realm of Native American legend or merely a classic campfire tale to frighten young, gullible children. I rarely watched sci-fi programs and never had read a single book on the subject. Ironically, as fate would have it, I would later become the topic of many such programs and books. I digress.

As a sergeant in Charlie Company (1249th Combat Engineers), it was business as usual as we headed up into the dense temperate rain forest of the Coast Range in Northwestern Oregon. The mission of Combat Engineering can be boiled down to two words: 'mobility' & 'counter-mobility'. In other words, ensure our troops overcome any obstacles (man-made or natural) and deny the enemy passage (or route them to where we want them) by placing obstacles in their path. On that particular day, our mission was to conduct training on private timberland near Saddle Mountain; just east of the coastal resort town of Seaside. We would be executing demolitions (explosives) operations at three rock quarries. Each site had a unique battle scenario to accomplish.

At the first site, we practiced cutting charges. This is where we would use plastic explosives (composition four or C4) to shear steel I-beams like a hot knife through butter in an effort to simulate dropping a bridge. Simultaneously, we also cut a five-foot diameter Douglas Fir tree in half by wrapping a belt of C4 around it. Both charges were a resounding success! One sheared steel I-beam looking like an exploded cartoon cigar and a whole lot-o-bark dust.

The second site held a 'complex obstacle' consisting of a field of surface-laid anti-tank mines followed by a triple-strand concertina wire fence. We were to clear a vehicle lane through both. In addition, we were tasked to construct a field-expedient 'claymore' anti-personnel mine out of a #10 coffee can with improvised shrapnel. After securing the area and checking for subterranean mines, we strung a 'ring-main' (a circuit of detonation or DET cord) through the mine field. DET cord looks similar to fuse cord with the exception that it contains a tremendously explosive compound (PETN or Pentrite) which burns at a consistent rate of 8,000 meters per second. It is said that if you could string a line of DET cord from LA to New York, it would take approximately 14 minutes to get to the other end! It is essentially used to synchronize several explosive devices to detonate virtually simultaneously. You definitely do NOT want to confuse DET cord with fuse cord! In any event, my squad set about placing C4 charges next to each of the anti-tank mines and tying them into the ring-main while another squad began fashioning a field-expedient (read: homemade) version of a 'Bangalore Torpedo' to breach the razor-wire obstacle. Normally a Bangalore Torpedo is essentially a 3″ plastic pipe filed with C4. Sections of this pipe are generally fitted together to form a pipe long enough to breach the entire obstacle. In this case we had to sandwich C4 between sections of U-channel fence pickets then wrap them together with duct tape (same effect). The homemade 'Bangalore' was then tied into the ring-main. Lastly, we constructed our field-expedient claymore mine by poking a hole in the bottom of a #10 can and inserted a blasting cap. Next, we lined the bottom of the can with about 2.5lbs of C4 then covered it with three layers of cardboard for wadding. Finally, we loaded the can with rocks, bolts, nuts and anything else that would ruin the 'enemy's' day. The can was then buried into a hillside (pointed towards the enemy) and angled about 12 degrees off the ground then it too was tied into the ring-main.

While I had not yet seen these creatures, there was a brief incident which, in retrospect, made me think they may have seen us. While I was directing my squad to emplace their charges next to the anti-tank mines, there was a rather loud, crescendoing 'WHOOOOP!' that emanated from the west end of the mine field. At that moment, I was bent over placing my own charge. Upon hearing this somewhat shrill noise, I immediately stood up and sought out the perpetrator as we were under orders to practice noise discipline during the exercise (in case the 'enemy' were nearby). As I glanced around the mine field, I was surprised to find all of my men still busily preparing their changes and not, as I suspected, goofing off. I shrugged my shoulders and went back to work. In hindsight, it seemed to me that the WHOOOOP sound had come from farther back in the tree line. But that made no sense as everyone was present and accounted for.

Once all of the charges were set and the area was cleared, I yelled "FIRE IN THE HOLE!" then pulled the dual-primed M-60 fuse igniters. The fuses hissed and began snaking their way towards their primary charge while we mounted up and began to convoy down to the safety staging area to await the 'report' of the explosion a short eight minutes away.

At that moment, we developed radio problems. The field commander could not reach the base commander back at Camp Rilea. I was tasked to take my HMMVE ('Humvee') up to the top of a nearby hill, where we had a 'two-niner-two' radio relay station set up, to see what the problem was. Upon arrival, they had already repaired the relay, so I decided to watch the impending explosion from that vantage point. Even from two miles away, the sight of 200lbs of C4 detonating is an awesome sight. The huge flash was followed by an even bigger black cloud which began to build into a mushroom cloud. Simultaneously you could see the trees in the immediate vicinity shudder in succession as a shock wave rolled across the forest below in a perfect concentrical ring. Finally, about a two seconds later, we heard the BOOM! Another resounding success.

The third and last training area was situated in yet another gravel quarry on a hillside that overlooked the second blast site. Here our mission was to emplace a 'cratering charge'.

As the name implies, this type of operation involves the making of a rather large hole. Generally, this is done to sever a road thus denying the enemy use thereof. To the uninitiated, an explosion is an explosion. To those of us who deal in the science of explosives, there are very distinct differences based upon the target, its composition, type of explosive (dynamite, C3, C4, ammonium nitrate, PETN, TNT, RDX, etc.), amount of explosive, its placement, shape of the charge, tamping, etc. Whereas C4 produces a super-hot/fast explosion, ammonium nitrate (essentially refined chicken or pig manure) soaked in diesel fuel for several hours, results in a 'slow' concussive blast. Properly placed and tamped, it will effortlessly relocate a generous section of real estate. It should be noted that, absent a standard issue shaped charge, we had the 'heavy junk' (read: heavy equipment) section pre-dig a starter hole with a backhoe. After emplacing several bags of diesel-soaked ammonium nitrate into the aforementioned hole, we (read: privates) filled it in and tap danced on it to tamp (pack) the charge. Once again, the area was cleared, and I initiated the dual-primed M-60 fuse igniters. I took my place in the waiting convoy and, per S.O.P., we began the descent down to the safety staging area.

Being a squad leader, I had the privilege of having my own Hummer, complete with a driver and an A (alternate) driver. Ours was the second vehicle of a five-vehicle convoy (2 Humvees up front, 2 covered troop carriers called 'deuce and a halfs and the Commander's Humvee in the rear). I took up a position behind the driver's seat and, as we were descending the narrow winding road down towards the staging area, I had the opportunity to enjoy the scenery. As an avid hunter, it is just second nature to me to spot for wildlife. As it was a rare sunny day in April, I had my window unzipped for a better view. Rounding a corner, I had a good view of the rock quarry where we had done our second blast at less than an hour earlier. Standing right out in the open, in the middle of the gravel pit, were three, jet-black, bipedal creatures. They stood in line, shoulder to shoulder staring directly at our convoy as it descended the hillside across from them. Between us was a ravine populated with eight-to twelve-year-old Douglas Fir and hemlock 'reprod'. At a distance of several hundred meters, I could not make out facial features or gender, but there was no doubt what I was looking at were not humans. Had these creatures been standing in front of a backdrop of trees, I most likely would not have seen them at all. But in this case, there stood three dark black figures contrasted against a light grey cliff of basalt on a bright sunny day.

In the middle stood, what I assumed to be, the alpha male of the group; as it towered a full head above the two creatures that flanked it. I would estimate it to have stood approximately nine feet high, with the flanking creatures approaching seven feet in height. Their silhouette was unique in that their heads sat directly on their shoulders with no visible neck. They all displayed broad, square shoulders and barreled chests which tapered down to a svelte waistline, unlike the creature seen in the Patterson-Gimlin film of 1967 (for the record, I am of the impression that the PG creature was either pregnant or had recently been so; accounting for her girth). The arms of these beings hung well past their knees. In the case of the two flanking creatures, they were exhibiting a swaying motion (rocking side-to-side) as the larger creature stood as still as a statue. Bear in mind that, all the while I was staring at the creatures, we were bounding down a dirt road with the occasional hedge of blackberry and Scotch bloom obscuring my view. That being said, I had approximately 25 seconds of viewing time.

At this point most people ask me, "Didn't anybody else see them?" "Why didn't you say something to your driver(s)? or "Why didn't stop your vehicle?

The answer is that…

I assumed I alone had seen them;

I was still in shock and disbelief;

my jeep didn't have a radio to call for a stop

and even if it did, we had a rather large BOMB ticking off behind us!

Once the vehicle rounded a sharp corner, I knew I had seen the last of them. I fell back into my seat with a mixture of shock for what I had witnessed and an odd sense of depression. It's a hard to explain what goes through one's mind in such a moment. Fate had somehow come together to create a once-in-a-lifetime moment that was now lost as suddenly as it had found.

My head began to swim with questions.

"Oh my God! They DO exist! And not just a solitary beast, but a group of them!

How could they exist and not be 'discovered'?

Having extensively hunting this area, how could they exist and I not have seen them, or signs of them, before? Some hunter!

What do I do now?"

I felt the sudden urge to tell someone, but who? I had seen something scientifically, if not historically, important and SOMEONE should be notified! There must be an authority that NEEDS this information!

I began to make a mental checklist.

The US Fish & Wildlife? No.

The Forest Service? No.

The zoo? No.

The police? HELL NO!

Then WHO?!?! And better yet, who would believe me anyway?

In the final analysis, I reluctantly decided (like most people do) to keep my mouth shut. Here I was a family man, a vice-president of a shipping company, and a Non-Commissioned Officer in the Army National Guard. I had worked long and hard for my reputation. And yet, with one simple sentence, "I saw Bigfoot!", I could throw it all away. Nope. If I knew what was good for me, I could never tell a soul.

Thus is the curse of the Bigfoot: living with the burden of the truth. A truth so absolutely incredible that merely suggesting that you 'might' have seen something that 'may' have been a Bigfoot will cause people to question your very sanity and even destroy your reputation. Great! I have jokingly suggested that I should start a Bigfoot Support Group for those afflicted with 'the curse'. One thing I can say, from years of interviewing other eyewitnesses, that there is something therapeutic in sharing such mutual experiences.

Arriving at the staging area, I immediately jogged back up the road in a futile effort to get one more look at these amazing creatures. Unfortunately, there was a knoll which blocked my view of the gravel pit. Again, an odd sense of depression swept over me. I felt a genuine sense of loss that was difficult to explain.

My activity hadn't gone unnoticed. Suddenly I heard footsteps heading my way. Then a voice yelled out, "Hey Neiss!" I turned and saw SGT Jeff Martin heading my direction. As he approached me, he looked over his shoulder to see if he was being followed. Satisfied that we were alone, he said something that I will never forget. He took a long drag off of his cigarette, exhaled, looked me squarely in the eyes and said, "I don't suppose you saw what I saw back at the second blast site?" It was more of a statement than a question. I could tell from the look in his eyes that he knew something. I felt overwhelmed at the possibility but decided to err on the side of caution. I replied, "I don't know Jeff, what did YOU see?" Once again, he looked left then right to make certain of our privacy and then stated rather matter-of-factly, "I saw three, huge, hair-covered, for lack of a better word 'BIGFEET'.

Trying to contain my excitement I hissed, "Yesssss! I saw them too!"

I was overwhelmed with sense of utter relief! I wasn't alone!! It wasn't that I needed validation of what I had seen. Corroboration could not have altered the truth, but it sure felt good. It felt somehow liberating. At that we began to compare notes.

Fate was busy that day. What were the odds? The odds that I ever would have seen them in the first place? The odds that someone else did (independently) as well? The odds that they would have even imagined that I had shared their experience and even in so considering would have had the courage to ask me? I can only guess that observing me looking in the direction of the quarry and straining to get a glimpse of 'something' was enough to pique his curiosity. Thank God! And finally, what were the odds that we were the only two witnesses? As I would come to later learn, we weren't!

That evening, we had an 'open post' so I decided to stay the night at the home of my friend (and Platoon Sergeant) Don Braden and his wife Lena. After debating whether to tell them my story, I reluctantly opened up (few beers didn't hurt either). Being my first 'Bigfoot confession', I found out the hard way that even your best friends can be hard to convert. After some initial ridicule, I had to settle for a bit of patronizing sympathy. This wasn't to be my last bout with ridicule.

Fate wasn't quite done yet. At the next Guard drill, Lena Braden and the other wives and girlfriends of soldiers were conducting a bake sale in the foyer of the armory at Camp Rilea as was their custom. I, on the other hand, was split-training in Portland some 100 miles east. Lena was in the middle of sharing my 'Bigfoot confession' with the other ladies when two soldiers entered the building. As they passed the bake sale table, they happened to overhear Lena describing my encounter, then froze in their tracks. They turned to her and asked her to repeat the story then both confessed to having seen the creatures as well!

A lot has transpired since that fateful day. It has been my personal mission to provide irrefutable evidence of these amazing creatures existence in an effort to gain their recognition and, if need be, play a role in their protection. I have spent countless days and nights conducting field research (including six major expeditions) throughout the Pacific Northwest, California and one foray into the

Mazetzal Wilderness of Arizona. My quest has given me the privilege to meet and/or work with some of the top researchers in the field: Peter Byrne, the late Rene Dahinden, John Green, the late Professor Grover Krantz (a.k.a. 'The Four Horsemen'), Bob Gimlin, Loren Coleman, Don Keating, the late Richard Greenwell, Larry Lund, Dr. Jeff Meldrum, Cliff Crook, Chris Murphy, Dan Perez, Joe Beelart, Ray Crowe, Dr. Wolf Henner Farenbach, Rick Noll, Thom Powell, Cliff Olson and the late Fred Bradshaw to name just a few. I thank them all for their generous insight, advice and companionship. Over the past 14 years, I have had the honor of appearing on more than nineteen television programs, several radio talk shows, and given speeches at numerous symposiums and colleges. Bigfooting will always be a part of my life and I look forward to many more adventures in the future

[some who read this might ask themselves why I included some much seemingly extraneous information. The answer is two-fold. First... since that fateful day in 1993, I had never written down a fully detailed account of my sighting. Second... while the events of that day are still quite fresh in my mind, it has been fifteen years and there is no guarantee that I will always be so cognitive. Last... in my years of interviewing eye-witnesses, I have always regarded the more detailed accounts as the most credible. I want to know what preceded and what followed the encounter. If someone allegedly took a photo or video of a Bigfoot, then surely there would be photos or video that preceded it (if not, then it raises "red flags" for me). So, this is the reason that I felt it necessary to be as concise as I can be. I appreciate your indulgence

Click

Kevin Carney & Sandy Nelson

5/17/18: Kevin had gone to bed earlier than I because he was going to get up at three am for a "wee hours of the morning" adventure with some other campers. After a hellacious drumming session, I actually joined him around 10:45 pm or so. I had just

fallen asleep when I got an elbow jab in my ribs which aroused me slightly from my slumber. By the second jab, I moved my body away from him and was about to say something – being annoyed - when I heard clicks from outside the tent. He was facing the back of the tent, and could feel an overwhelming feeling of energy.

132

I stated quietly, "Clicks." Kevin acknowledged. We started whispering… He reported hearing four knocks behind the tent at a high level from the ground… five feet or so… "with the cadence of a second hand…" Then three rock clacks at a lower height off the ground… three feet or so and closer to the tent.

A few minutes later there was what seemed like random rock clicks… small ones… that sounded like someone playing with my gifts I had left down the trail behind us. As the individual(s) approached our tent, it was closer to the ground. Kevin had heard the clicking sound and woke me up. I heard it right away. He thought there may have been two individuals, but wasn't certain. There were some bolder clicks and others that were softer. Kevin said it was like they could see each other, but couldn't because of the darkness and the tent wall. But they seemingly could "see" each other. I must note that we sleep on a four-inch memory foam pad placed on the tent floor.

Kevin then turned onto his back and we held each other's hand. As we were lying there intently listening for any movement outside, Kevin whispered that he felt a warm wave of energy wash over him from his ear lobes to his ankles… like petting a sleeping dog. He could feel black energy flow through him twice, and then over his lower legs once again. The black energy was described as a positive feeling and flowed through his body except for his head, feet, and approximately the last fraction of an inch that touched our mattress. Kevin stated he felt immense power, alertness, and awareness.

Outwardly, he asked about them about checking his eyes. He was telling me that he was seeing a light, like a morning fog, coming from within his eyes outward. It was mostly on the right side. He wasn't fearful, but very calm. Was he scanned?

Kevin whispered as he saw twinkle lights on the inside of the tent, between us and the ceiling. They were greyish white with yellow sparkles over me. At this point, I turned onto my back.

The individual(s) seemed right outside the back of the tent nearest Kevin. We continued to whisper to each other. "What should I do?" whispered Kevin.

My calm and assured response was, "Talk to them." It was interestingly to me, there was no fear… period. This experience seemed very natural. Although I had no idea exactly who was outside the tent, I knew bears, cougars, elk, and deer didn't click, so whatever it was wasn't going to hurt me. Besides, Kevin was with me, so I felt completely safe.

There had been excited clicking, possibly between two individuals, when I first heard them approach, but then it settled down to one individual as Kevin started to speak quietly. "I hear you. I hear your clicking… I like your clicking," The clicks continued intermittently.

We could hear movement outside the tent at our feet. The tent was dark and the movement seemed to go from our left to our right, toward the front of the tent. The individual didn't leave, but continued to click. Kevin was very calm, using a soft low reassuring voice. We were very alert in anticipation of what might happen next, not knowing the extent of this interaction.

The individual had repositioned itself on the side of the tent above our heads.

Again, Kevin expressed his gratitude for the clicking sound, and the individual clicked in response. The clicking was like a crisp snapping of a tongue on the palette of your mouth. The clicking happened a couple of times before Kevin asked, "What is your name?" Silence.
"I really like your clicks; can I call you Click?" The individual clicked once.
"I can call you Click?" Click…
"Sandy is with me. Do you know Sandy?" Silence
Kevin continued as if he knew this was a Forest Person…
"Are you a boy?" Click…
"Are you a girl?" Silence
"So you are a boy?" Click…
"How old are you?" Click, click… a slight pause and another click as Kevin said, "so you are two… or three?" Click, click…
"You are a boy and you are two?" Click…

"Do you have a brother?" Click…
"Do you have a sister?" Silence
"So you have a brother?" Click…

"Do you like apples?" Click…
"Would you like an apple?" Click…
"Okay, we will leave an apple for you in my hammock."

We whispered to each other as we began to understand that one click was a yes… silence was no… and a pause or hesitation with a weaker click meant possibly or probably, as if it were thinking about the question before responding OR perhaps it was uncertain of the question content.   I suggested we stick to yes and no questions. The clicks were very clear, precise, & crisp.   They were not hollow or mucousy at all. Some were softer in tone or volume when he was possibly uncertain or hesitating.

"Would you like us to bring an apple for your brother?" Click…
"Should we leave the apples tomorrow?" Silence
"Should we leave the apples before morning?" Click…
"Before the other campers get up?" Click…
"Okay, we will leave apples for you and your brother in the hammock."

"I can't see you through the tent. Can you see me through the tent?" Silence

There were several times when a dull white light would shine on the roof of the tent… like someone was coming up the trail, but we couldn't hear anyone pass by the tent. It didn't show the tree limb shadows that would have been above or beside the tent. It was odd.

Kevin continued in a calm reassuring voice…
"I'm going to put my hand on the tent fabric if you want to touch me." Kevin explained each step reassuring Click. "If you want to touch me, you can. I'm going to place my hand on the tent now. I've got my hand on the tent, so if you

want to touch my hand, you can." We waited… Click did not touch Kevin's hand or fingers, but pushed the tent fabric between two of his fingers. Then, a second quick poke. Kevin giggled with delight. "AMAZING!" Kevin excitedly whispered, "He's right there." Indicating the individual was right above our heads… mere inches away.

Approximately one and one-half hours into this interaction, we saw a light shine repeatedly from the direction of the road. We could see shadows of tree limbs on the tent over the door. Click would be very quiet during these times. I whispered to Kevin, "Who in the hell is shining that light into the forest?" It was frustrating…we didn't want Click to get scared and leave, plus our campers knew better than that!

After a brief discussion between us, Kevin explained that he was going to get up, put his boots on, unzip the tent and he made the sound of a zipper. He explained each step, reassuring Click that he would not look for him, but he wanted to see who was shining that light. He stated that he wished Click would wait for him to come back. Kevin did exactly as he had said.

Once Kevin was gone, I felt very protective of Click. Speaking to him through my thoughts, I found myself telling him to hide if I saw the light shining in the direction of the tent, and to be very quiet… He was. I also told him we would not tell anyone of this encounter if he didn't want us too. Click was silent. In my mind, I already knew Kevin was probably telling other campers. After several minutes, Kevin returned to the tent.

Before Kevin got resettled, I informed him I needed to use the restroom. He suggested we do that before he got now and he would escort me. He explained to Click what was going to happen…we were both leaving the tent, but hoped he would wait for us to return. As I turned over on my side to get up, Kevin heard three distinct foot falls. As we exited the tent, he stated excitedly… "He's right here…I can feel his energy… He's right here." His hand was shaking with excitement as he led me down the trail and onto the road.

As we went by the campfire, a few campers were still up and wanted to ask questions, but we continued on to the restroom. "I knew it!" He told me!

On the way back, I saw an apple at a little boy's (one of the campers) gift area. I grabbed the apple, vowing to replace it in the morning and we hurriedly returned to the tent as we hoped Click would be waiting for us to return. Kevin placed the apple in the hammock and we returned to our bed. As we settled in, we waited to hear Click. Was he still there? After a few minutes, Kevin asked the question…

"Click, are you there?" Click…

We were excited…and the communication continued.

"We left an apple for you in the hammock. We will leave one for your brother in the early morning before the other campers get up." There were several excited clicks.

"Click, are you a Sasquatch?" Silence
"Are you a Forest Person?" Silence
"Are you an Alien?" Silence

"Are you a Bigfoot?"   Pause…then a softer Click…   (I started to chuckle as I could envision this little fellow looking at his feet, pondering the question of the size of his feet before he clicked.)

"So, you like to be called a Bigfoot?"   Click…

I'm not certain that he was responding to the name, but probably that we acknowledged the size of his feet and he was pleased.

"Do you look like me?"   Silence

"Do I look like you?"   Silence

"Have you talked to a person like me before?"   Click…

"Have you talked to a person like Sandy before?"   Silence

"So, you have talked to a man before."   Click…

"Do you live here?"   Click…

"In this valley?"   Click…

"Even when it snows?   Click…

"Do you like the snow?"   Pause…then a softer Click…

"So you have talked to a man like me in this valley?" Click…

"Do you live in a cave?"   Silence

"Do you live in a tree?"   Silence

Kevin stated, "I can't see you through the tent. Can you see me?   Silence

"I can't see you during the night, can I see you during the day?"   Click…

"Can Sandy see you?"   Silence. I sighed heavily, but got the impression Click was shy. Kevin continued…

"Will you bring your Mother?"   Silence

"Will you bring your Father?"   Silence

"Will you bring a friend?"   Silence

"Will you come back tomorrow?"   Click

"I look forward to visiting with you tomorrow night," stated Kevin.

Interestingly, if I spoke out loud to Click, he would not respond, but if Kevin asked the same question, he would respond. He was definitely comfortable interacting with a male.

At approximately two am, Kevin stated we were getting tired and needed to go to sleep. Click acknowledged with 1 click. We could hear him move off down the hill from our heads. He clicked as he went…two more clicks getting more and more distant.

Once Click left, little sparkles filled the tent again…some with little trails of light. I put my glasses on to be certain I was seeing these… I was. Then it went dark.

What an amazing experience! I had never heard of anyone having this type of communication exchange. This was an intelligent exchange. It understood what was asked and gave the same affirming answers when questions were repeated.

Kevin made the decision to use his headlamp and started journaling the interaction while I went to sleep. Unannounced to me, Kevin got up before daylight (4:30 am) and went to retrieve two more apples, placing them on a nearby stump as promised.

Ultimately, he placed the apples on the stump next to the hammock because the apples were getting lost in the folds of the hammock. Unfortunately, the apples remained there for the duration of the campout and Click never returned to visit or get the apples.

The next morning, a nine-inch foot print was found near the corner of the tent, just off the entry tarp, toward our heads. Two more prints were found leading down the hillside.

Before breakfast, I suggested that Kevin spend some time in the forest by himself, giving himself and Click the opportunity to see each other. When he returned, he had not had a visual sighting, but had found some prints he wanted to show me.

After breakfast, he and I slipped away. We bush whacked… finding possible glyphs along the way along with an animal's bedding area. We finally emerged at a little pond of water. There were two prints in the water…a right and left foot. Narrow (female), and you could see the outline of toes in both of them. I purposely didn't take my camera, because IF there was a possibility to interact again with Click, we wanted to do so and felt a camera might prove detrimental to that end. We proceeded over the ponding water to a lovely forest of cottonwood trees… Beautiful!

We proceeded a little further and circled around to cross the pond again… BUT… Just before I stepped over a downed tree, there was a magnificent print.   It was the second largest I had ever seen. The toes were evident and the little toe had pushed up mud. Okay…now we HAD to go back and get my camera and a measuring tape. I marked the print with a Y stick I found and a directional stick placed in the Y. Kevin placed an apple up on a limb in a tree.

When we got back to camp, it was just after two pm. Barb (Shupe) was waiting on other hikers who had expressed an interest in hiking to some lakes, but because it was getting late, she didn't feel there would be enough time before she had to leave to pick up Thom Cantrall. I asked if she would like to come for a walk with Kevin and me, explaining the prints. She was very interested and after a quick snack, we were off, bush whacking and retracing our steps.

The footprints were videoed and measured. It was fourteen-and one-half inches long… Impressive! Interestingly, in the short amount of time that we were gone, the Y stick had been pulled out of the dirt and was leaning on its side. The apple was still on the tree limb. At one point, Kevin had strayed away from Barb and me. He had heard one click just before I called to him…oops.

We left Kevin in the forest and returned to camp. Unfortunately, there was no further interaction or sighting of Click. On the way back to camp, Barb found what appeared to be some knuckle prints. We both curled our fingers back and tried to spread our fingers apart to match our second joints with those of the print. They were just a bit bigger than ours. These were upstream from the marshy area where Kevin had found the two prints.

As Kevin and I reminisced about this experience, I couldn't help but feel a little uneasy at the thought of this little fellow returning to his Mommy, asking her what a Sasquatch, or Forest Person, or an Alien was. He probably got a lecture about interacting with the hairless people… Sorry Click…

Since the campout, I have reached out to a few people who are reliable, knowledgeable and trustworthy regarding Sasquatch.

Arla Collett, from Oklahoma, has the Forest People living alongside her on her property. She stated she has heard this clicking many times. It is one of the first types of communication used between a mother and infant Sasquatch.

All in all, our interaction with Click was incredible and we are thankful for this experience.

Dale Boswell

I still cannot believe this Creature was in my backyard, standing a few feet from me when, unknowingly, I captured its face on film from the chin up. I believe it was a mischievous juvenile, judging by the expression of surprise on its face as it glanced directly, turned and wobbled off.

The being was totally silent, very fast and practically invisible to the naked eye. That is why I didn't see it at the time. I downloaded the film clip and basically forgot about it, then, in March of 2021, I was in the process of deleting some old files from my desktop PC when I decided to take a closer look at a particular video. What I saw was amazing and unbelievable. The film showed at least four different, moving anomalies.

Since rediscovering the file, I have studied and analyzed the fifty-eight second film clip extensively… frame by frame. I believe the "green fog" surrounding the creature is its hair.

In the original footage, the "Glow" appears to be the shade of a green darkness… and it blends perfectly with the night. This indicates to me the fact that their hair is part of the secret to disappearing act. They possess an uncanny ability to distort and refract light, therefore blending into the surrounding environment, whether it be day or night. Using my professional graphics software skills, I have worked, very hard, to present my findings with as much detail and information as possible. Of course, I do realize and understand that until the species is scientifically recognized, no amount of media and/or photographic evidence is going to be taken seriously by the general public.

What Happened

On May 18th 2018, around 5:30 in the afternoon, I was perched on my homemade gazebo-like deck that I like to refer to as my "Booger- Hut". I had just purchased a new cellphone and I was checking out the video options. I wasn't going to film anything in particular. Fortunately, I had my finger about an inch from the red Record button, when suddenly, I heard a loud, crashing, limb-breaking, rustling noise coming from the woods to my right. Immediately,

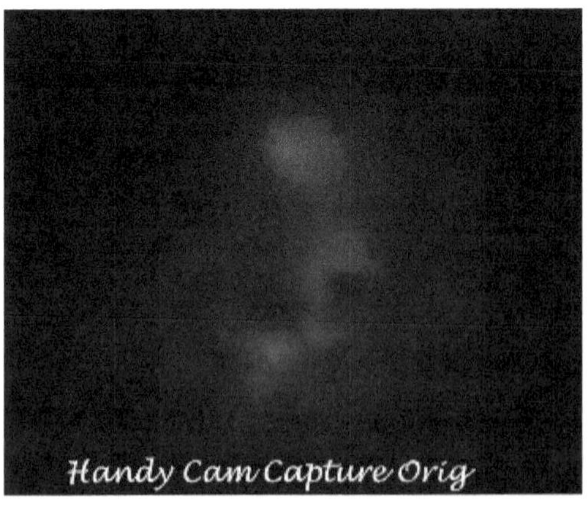

with the phone in my left hand, I hit 'Record' and twisted my upper body towards the area of commotion.

Initially, I thought I saw the blur of a tree

limb springing back up after another limb had fallen and hit it on the way down. Still, not wanting to miss any footage should there be any live critters, I held the phone as still as possible while continuously scanning the area with my naked eyes. I didn't see any movement, and it got relatively quiet. I filmed for almost three minutes straight. The only time I turned the camera away from the tree line was to climb down from the deck. I walked over to the area to investigate while still filming. I didn't see any sign of what had made the noises.

There were old limbs and branches on the ground, but nothing new, as far as I could tell. If it was an animal, it was really hunkered down and hiding, so, I walked back to my Booger Hut, sat down, and reviewed the film clip.

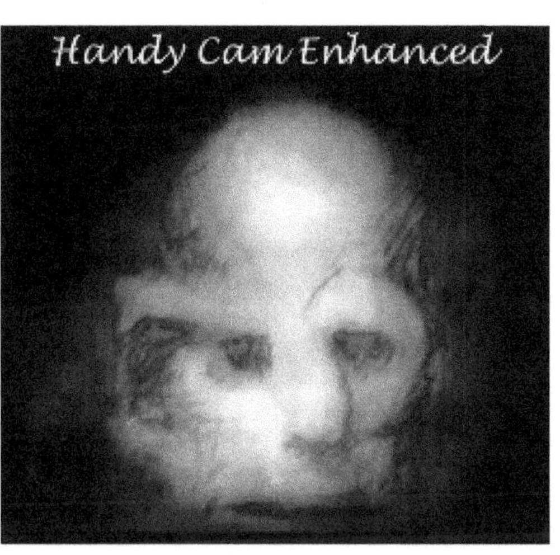

A BABY BIGFOOT? I watched it over, and over, and over, and over. I couldn't believe it. The "blur", and the loud noise, that I initially thought was a tree limb looks like a freaking monkey! As far as I know, we don't have monkeys running wild in Alabama.

Shane Honea

My interest in the creature known as Bigfoot started when I was just a child. I was maybe eight or nine years old when I saw a late-night movie called "The Legend of Boggy Creek". This movie held my attention more than any other movie I had ever seen. I was so captivated by this that I really started developing an interest in the Bigfoot field. After that, I started watching every movie, documentary and reading every book I could find on the subject. I did this for many years and still do so today.

When I was about sixteen or seventeen years of age, I had my first encounter with this creature. It happened when I was coon hunting with two older friends in Barrow County, Georgia. At this time, I had been hunting on this 3000-acre tract of land for about two months.

It was about 9 o'clock in the evening when the incident took place. It was fairly cold that night and very dark in those woods. We had smelled a terrible odor which, at first, we attributed to a skunk. After we passed through the area and out of the area of the odor, we began to hear movement in the brush off of the trail behind us. We kept moving deeper into the woods and still some distance behind the hounds which had, by this time, struck a trail and were in pursuit of a raccoon.

The noises behind us kept getting closer and louder. We had first thought it was one of the dogs that had come back into us for some reason. This was unusual to us as all of the hounds we had were champion coon hounds and would rather die than give up on running a raccoon. We began to curse and swear as we thought it was one of our hounds, but we soon discovered that it was not one of our dogs as we could plainly hear each dog's voice in front of us as they were bawling away in pursuit of the raccoon.

My friend Hubert jokingly said it must be a werewolf that was following us. He told Ray and myself to turn off our hunting lights and wait for a signal from him. At his signal, we were to point our lights in the direction of the noise moving ever closer to us. Hubert had a .22 rifle that he carried for shooting the coon out of the tree after our dogs treed. Hubert made the comment that if he didn't like what he saw in the beams of our lights that he was going to start shooting.

At the signal Ray, and I turned the lights on high and aimed the beams in the direction of the noises there... In the light was a tall, dark figure that ran off as soon as the lights were on it. It was huge and ran upright on two legs just like a man. this shocked all three of us and Hubert never fired a shot as he was not sure what he had just seen!

This was not a man nor was it a bear as it was too large and bulky to be a man and two large to be a black bear. Besides, a bear only walks on his hind legs for a short distance, and awkwardly at that. This thing was sprinting away rapidly while leaning slightly forward. As it moved, it was breaking branches and tree limbs as it ran up a wooded hillside and was soon out of hearing distance.

We did not care to investigate it as we were too scared and shocked to do so. After we calmed down, we made our way to get the dogs and get back to the truck. After putting the dogs in the dog box and stowing our gear, we were flying down the old logging road getting out of there!

On the way home, Ray and Hubert began to talk more about the experience as well as the footprints found on this track of land.

I continued to hunt there for many years afterward but I never again saw or heard anything again. We did continue to smell that same odor that we smelled when we had the encounter. It always unnerved me a little every time we hunted in that area.

Was what we saw a bigfoot? I think so and will continue to believe so till my dying day! Since that first encounter, I have continued to study the bigfoot creature. I have had no further sightings, but have found foot prints as well as other evidence since that night coon hunting in Barrow County, Georgia. Will I ever see the creature again? I hope so. I continue to investigate the woods in my area and follow all the leads that I get about this creature. Someday, maybe, the mystery will be solved. Until then I will continue to search for this creature in the hopes of making contact.

I have told this story to only a few people over the years. Some believe me and others only laugh. Some folks don't believe in such things, but I do as I have had a glimpse of one and I have come across more evidence in my years of amateur research.

Who knows what dwells in the vast forests of North America? The silverback gorilla was only thought to be a legend until science proved otherwise only recently. I await the day When science proves the existence of bigfoot then and only then will everyone believe the realness of this creature.

Shane

My Meeting with the Big Guy

Keith Bearden

I was standing in front of the Big Guy... Arla Williams was beside me. Behind me there were four more of my friends as well as my wife. They all witnessed something very, very special.

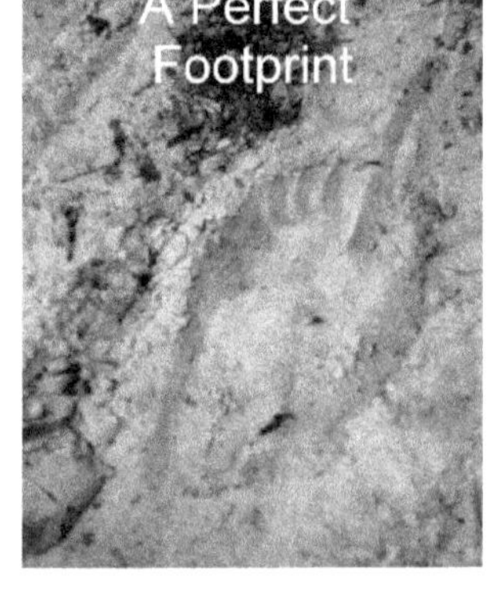

It was late at night and, as such, was dark, except for a moon that was shining fairly brightly, but wasn't directly overhead. It was casting long shadows due to the angle the light was penetrating the forest canopy.

We all heard him moving down the hill towards the horse trail we were on. Suddenly, he stepped out in full view, facing Arla and me. Arla was, and still is, my mentor and was the person who had told me I would meet him. That had happened a few months prior.

Now, she told me to walk toward him as he had things to tell me… I did. He was just feet from me, standing in the trail and very visible in the soft, filtered moonlight. I had moved less than a few feet when all the hair on my arms stood up. I felt a vibration that started inside by my body and continued as I heard audible words in

my head that simply told me quickly "I am he whom you

sought to help teach you our ways and to gain understanding for the good of yourself and to help show others.

"This is extremely hard to explain to others, but after the first few words which I had audibly heard, it all blended into simply visuals and a very deep knowing of them, their people and their reason for being on earth. The entire "presentation" was like a Power Point presentation where the audio was coming directly into my mind while a video presentation was playing and showing a running scene in the background. It then accelerated into scenes, vignettes of them doing things, performing tasks and moving in super-fast motion, much like fast forwarding a movie.

This seemingly lasted for hours. I felt dizzy and was a little nauseated and getting very tired. All of a sudden, absolutely everything stopped… He turned slowly and simply walked off the trail and stood there. I took a few steps back then retraced my steps back to where Arla and the others waited. As she looked at me, she knew that which I had experienced and smiled at me.

I walked back to the group and stepped up beside my wife. I was totally drained. It was then Cathy walked towards him a short way, but not as near as I had approached him. I watched as her left leg slowly lifted from the ground and was moving straight in front of her at a near forty-five-degree angle. That pose lasted only a few seconds before she put the leg down. She didn't cause that halt in her advance, he did. She received no message. It is important to note here, she had metal in the leg that lifted left over from a past auto accident.

An interesting side note on what happened here. While I was being "downloaded", Arla said it appeared that I grew to a height of ten feet tall… that I looked very close to his height. Cathy said that, to her, I completely disappeared. She didn't see me until I stepped back to the group.

Here is what I know, and it helps to explain, at least in part, the things they can do like disappearing. I was surrounded with a vortex of energy. I very definitely could tell that right away. That same energy was still there, but much less force when Cathy walked forward and her leg lifted. It was like the metal in her leg was being attracted to a magnet.

I have, since that night, been able to KNOW things at certain times about them. Sometimes knowledge just comes into my head randomly. I dream of things about them at times. I can be asked a question about them or hear someone say something and I know the exact answer or can see and experience what others do. It was extremely humbling and has changed my entire life. That in itself is a completely different story and enough to fill another book.

If anyone is at the stage that they have established contact, they should be patient and if the Sasquatch people have decided to share more, it will come when and only if they and the receiver are ready. It won't be anything anyone would have ever expected. Fear, cannot be a part of things, as they will only reveal what they think you are ready for.

The one thing I have learned more than anything is, they will teach one more about himself than he ever dreamed possible.

Scott's Story

The following narrative is a second accounting of the incident with Keith Bearden in Georgia. This comes from a person new to our group who is the brother of one of our long-time members, Farrell.

In Scott's words…

I have just now gotten to the point where I would like to share my experience of last October (spring of following year). Keith has told the story, but I wanted to give my perspective. The details may not be spot on, but it is what I recall…

Last October (2014) was my first trip searching for the sasquatch people. I was behind my tent just looking and listening to the woods when Keith walked up with a set of parabolic ears. He said when he walked up, that he felt there were bigfoot around. With the parabolic, I could hear movement deep in the woods.

Keith left me there alone and I tried to communicate. I let them know I was respectfully there, who I was, why I was there and let them know I was without a camera. By the time Keith came back, it was dark and he brought three others, Arla, Cathy and Gail.

When Arla first arrived, she said she sensed multiple bigfoot. She talked to them as a mother would talk to a child… sweet, soothing and constant.

We stood in one spot when they started whispering: "Eyeshine at one o'clock, eleven o'clock, etc." I saw nothing. Keith pulled me in front of him and the forest came alive. I could see the eye shine. One set was low, as if they were lying down. Two sets looked as if they belonged to children, and the last set was about seven feet off the ground at a distance of maybe fifteen feet. It was dark and I could see nothing else, but others seemed to notice the blackness of shadows and subsequent movement.

At one time, Gail said there was a female at three o'clock. I saw nothing, but when I directed the parabolic, I could hear a constant, intermittent sound coming from her location. When I placed the parabolic in the direction of the two little ones, there was also a sound coming from them, but it was more of an energy sound, if that makes any sense. Only in those two spots would I hear those sounds. I was amazed and in awe.

We left as Keith took us to what is considered the area of his teacher, and of a particular clan of bigfoot people. As we approached, there was a small bridge. Keith stated that we would be greeted by two bigfoot people, one on each side and that they would escort us in.

As we were walking the path, I held the parabolic at each perpendicular side. Holy Crap! … but I could hear distinct footsteps crunching through the woods.

When we stopped, I would hear an extra step or two. They followed us down the path two hundred to three hundred yards. I could see eyes shining on the left and on the right within ten to fifteen feet of us. I would say there were at least six of them, including the two following us in.

Arla then said, "There is a big one at the end of the path."

The others could also detect him, because he enveloped any light at the end of the path. I asked Keith how big, he said at least nine feet, and it was his teacher. Arla pushed further while we stayed. She stopped half way between us and him. She called Keith down where they stayed for five to ten minutes. The moment Arla started forward down path, I directed the parabolic in the direction of the end of the path, and heard the same energy sound I had earlier with the two children. After Keith and Arla returned, the sound stopped.

We walked back and I'll damned if those two didn't follow us to the bridge. I heard heavy footsteps the entire time.

I was blown away and forever changed. Honestly, what I heard and saw had a drastic effect on me for weeks.

A Blue Mountain Affair

July, 2013

Michael Beers

It was summer and a group of us were on an adventure in the Blue Mountains in the arid, desert country of southeastern Washington State. As part of this outing, we headed out one evening with a full moon and a clear night sky to ply our task as we knew it at the time. Our large group had split into smaller, splinter groups and had spread out along a major ridgeline near the U.S. Forest Service Guard Station at Godman Springs.

The spot my wife and I chose was in the saddle just down the west side about a quarter mile from

the structure. Off to our left was a team of two, one of whom was a well-known scientist. While getting ready to head to this location I asked what he thought of this Bigfoot stuff.

He gave a good belly laugh then paused and said, "I think you guys are crazy, but I came to see for myself what this is all about." His life changed forever on that night.

My wife and I stood in our spot paying close attention to the night noises and the other team on our left that was interacting with at least one being of some sort. It sounded as if the humans would knock twice, they would get two knocks back. If they knocked three times, they got three knocks back. One knock then one knock back, three knocks then three back.

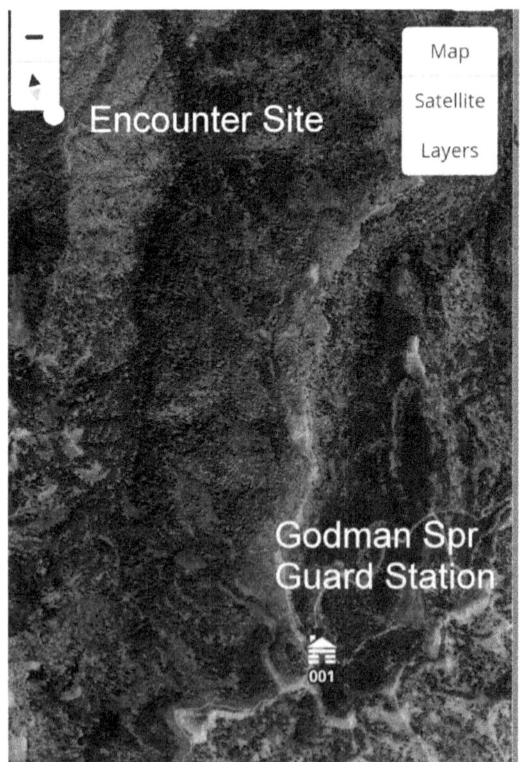

As we listened to all of that, I could hear movement below us and to the right of the standing burned out trees from a recent forest fire. That sound got closer and closer to those trees. As I said, it was a full moon, so we didn't need a flashlight to see the ground, the textures and the muted colors of our surroundings. We had just heard what sounded like some bark come loose from one of the trees and fall to the ground. I asked my wife if she was comfortable with me going down the hill to investigate the noise.

She answered that, yes, she was fine with that, so off I went to the bottom and stopped just inside the tree line at the point where two wind-fall trees intersected and almost made a perfect ninety-degree corner. Three burned out trees we're off to my right, two full sized trees and a shorter broken one. These trees were about ten feet from me.

I began talking to whatever was out there, if anything. I said into the forest, "Hello, my name is Mike and I came to visit with you. I paused for a moment then said, "Thank you for allowing me to be in your home." When I received no response, I said, "Hello, is anyone there?" Nothing came back to me.

The team to my left was making some noise, but I stayed focused on what I was doing. I then picked up a thin rotten stick and tapped lightly on one of the fallen trees in front of me and continued to do so for a few minutes. I waited and said, "Well, if nobody is there, then I am going back up the hill. Thank you for letting me be here."

As I turned to re-join my wife up the hill, suddenly, what I thought was a broken stub of a burned-out black tree started walking out from beside one of the taller black trees and towards me slightly.

At this point, I realized that it was a tall Sasquatch and it was now only about six feet from me. It was a mature, mildly robust, male, and he was just over nine feet tall. As he came out of the shadows toward me, he turned to my right and walked past me. I could see him well enough to verify that he was a male and I finally got a good enough look to see his ear size and placement. His hair was dark, short and worn from its sides. The skin was light colored like a light gray.

He paused and appeared to look me up and down, as I did the same to him. There was a moment of silence as we both took it all in, then I introduced myself and he began to walk away. He was in direct moonlight and I saw fantastic detail… muscle movements, expression, his eyes looking around and even a small amount of finger movement.

The other team, unaware of what was going on with me, yelled out again. He looked that way then turned and began to walk away from me downhill.

I said, "Don't leave, I came so far to meet you."

He took another two or three steps and then stopped on a flat area about thirty yards from me in the open. He then turned and took two steps backwards into the shadows.

I yelled out to the other team, "I have one over here and I'm looking directly at him."

That team then came to my location to see for themselves that which I was seeing. When they got to me, I asked what they had going as it sounded like the creature they had been interacting with was copying them. I was then told, no, that it was actually the other way around. That they were actually copying the banging and knocks that they were hearing.

I then explained what I had just experienced, and the two of them asked where, exactly, was that... where was the creature? I pointed it out, but they couldn't see it in the shadows until the male Sasquatch moved and they locked onto it. The scientist asked me if I was sure that is a bigfoot. He said that he couldn't be sure. I then asked if they wanted to stay here with the male and I would go over to where they had been... the site where they had their interaction going. They said they surely would, so I left them to have their own experiences with the large male below them

When I got to their location, I introduced myself and said thank you for allowing me to be in your home, as I always do. I then proceeded to tap a log lightly as a creature matched any and every combination of taps I could think of. This went on for quite some time, and I was having a fantastic interaction experience. However, I wanted to return to see what was happening with the other team.

I said into the dark forest, "Thank you for playing with me tonight and thank you for allowing me to be here."

I then turned and started to head back to my original spot. When about halfway there, I could hear some excitement in the voices of the other team. Just as I arrived, I asked, "How's it going over here?"

One of them said, "Fine, but the dynamics have changed…"

I asked, "What do you mean?"

He then explained that shortly after I left, the being had begun throwing little sticks at them. They said that it leaned forward partially into the moon beam and threw a stick that fell short of them.

We talked about what was happening and one of the guys said he didn't think it wanted them there, so they were going to leave, and did so.

I stayed and started to talk to the male Sasquatch. I thought it was all over because he wouldn't respond or throw sticks at me… he just stood there. I could see him perfectly as he was mostly in the moon beam with only one side of him in the shadow line. I asked him to join me out in the open, but he only shuffled

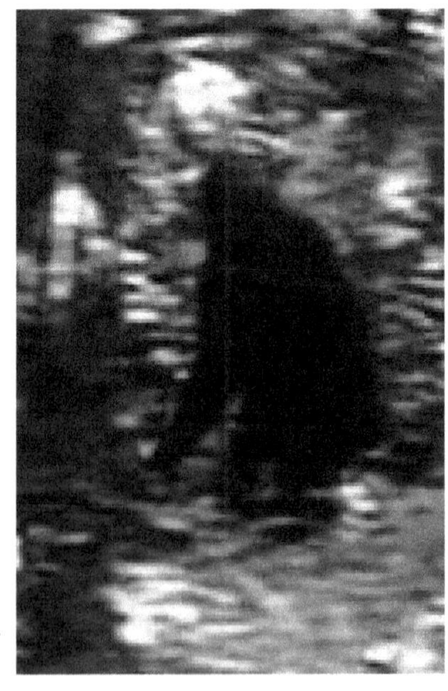

his feet and would not come to me. I thought that, perhaps, if I moved up past the tree line he might also do so. I begin to walk while talking softly about how beautiful their home was and that I had traveled a long distance to come visit him and the others. I had moved about ten yards and he still would not come with me, so I walked a little faster, still talking to him.

I got out of the tree line and about twenty-five yards up the hill in the open and found a rock to sit on. I sat on a good size rock that was only about six-inches above the ground level which was next to a taller rock that had a flat top. I continued to talk and asked if he would come sit with me. I repeated that question a few times and he started walking my way.

Oh, the excitement I was feeling at that moment. I thought for sure this was going to happen. He was getting closer and closer to the edge of the forest. I could see him clearly as he was now out in the open, but would still not commit to coming to me. I slowly stood up and proceeded to tell him, "I understand if you don't want to come closer but I was very happy to have met you." I explained I would do no harm as I was only here to meet him and his people and that is all. He suddenly took another couple of steps towards me and his eyes started to glow, getting brighter and brighter like a dimmer switch had been steadily turned up. It was a saturated medium light greenish yellow but mostly green. I said to him, "Wow, you have pretty eyes." Then the thought popped into my head that I could have said a hundred different things and I chose that. I seemed to have left my "What Do You Say to Sasquatch for Dummys" book at home.

He looked around while I was talking to him, then his eyes started getting dimmer and dimmer, until I could no longer see any light at all. He was back to what I will call "Normal," even though normal doesn't apply here. I thanked him for showing me that amazing and wonderful thing he just did. I explained how blessed I felt that he trusted me enough to show me that.

He then brought both hands up in front of himself at chest level and out away from his chest slightly.  He proceeded to bring his fingers together to a point like you would if you wanted to grab something with only your fingertips. He did this with both hands. Next, he brought his two hands together, keeping the fingertips in the same position. He then began twisting his hand back and forth with only the tips touching the whole time. This evolution lasted for about a minute to a minute and a half. He paused twice during that time but never separated the tips of his fingers until he was done.

I explained to him, "I don't understand... I want to but I don't. I asked him to show me again but he did not. I asked him to please say something, but he did not. I asked him to please come and sit with me as I brushed the dirt off the taller rock next to me. I've then sat back down on the rock I had been sitting on earlier and continued to talk to him and repeated the invitation to him to sit with me.

After nothing more happened besides him shifting his weight now and then, I slowly stood up and explained that if he wouldn't come sit with me then I was going back up the hill to my wife and I pointed up the hill. As I pointed, I looked up the hill to see my wife was no longer there. In fact, I found out later, she had left to re-join the bigger group, leaving me there alone with the big fellow. She had missed the whole event. She left when the other team came down the hill to my left and started being loud.

I explained to the big male that I was heading back up the hill now to find my wife. I thanked him more than a few times for the experience and for allowing me to be there. I invited him to our camp as it was on the same ridge, just a little south of this location. I then started walking up the hill, hoping he would come along behind me, but no, he turned his back to me and walked back into the dark forest and I never saw him look back. The whole event had lasted about fifteen to twenty minutes in duration, but it will be with me the rest of my life.

My first sighting was over forty years ago, when I shared a trail with a large, younger male in daylight at a distance of about twenty feet. To date I have seen eight, ranging in age from young too old, both males and females. None have been aggressive nor has there ever been a bad smell associated with them.

If I never see another, I would still consider myself very blessed. I still have not been able to find out what the hand twisting meant. I did, however, read a testimonial from a woman in Idaho that saw the same thing from a young Sasquatch.

No matter what path one takes, remember to look up once in a while, for the beauty and wonder of nature is everywhere. Please also remember that respect is everything and you get what you give.

Michael Beers

Voices on the Lake

Kyle's Story

August twenty-fifth dawned warm with a promise to get even warmer in this far, remote area of north-central Washington state. Taking advantage of a rare day free of the demands of a growing business, we were happily camped on our

site near the tiny town of Lincoln, Washington on the banks of Lake Roosevelt.

Lake Roosevelt stretches some one hundred and fifty miles behind the Grand Coulee Dam all the way into Canada. In fact, eighty eight percent of the area drained by the lake is in Canada. The land around our campsite is semi-arid with an annual average of about ten inches of precipitation.   Consequently, this area makes for a very special summertime recreation area. The waters are never more than about sixty degrees even in the heat of summer, so a quick swim is always a refreshing thing. The land surrounding the lake is rough, steep and rugged.

There are pines and firs growing in the shaded areas of the north facing slopes and in the deeper gorges where the glaciers left enough soil to support the growth of a tree, but, for the most part, rock predominates. As the illustration shows, massive rock scarps rise directly from the lake and form a formidable barrier to egress from the lake in most areas.

Much of this lake's shoreline further north from out site falls under the jurisdiction of the Federal Government but in our area, it is under local control. The east side of the lake is mostly under private ownership with large ranches reaching to the edge of the water in many cases. The west

side is all part of the Colville Indian Reservation and is managed by the tribes thereon.

Wildlife abounds here with mule deer, whitetail deer and bighorn sheep predominating. Recently the moose have returned to the area with cougar and bear being in ample supply here. Myriad smaller species are found everywhere. Of course, waterfowl abound here as do both bald and golden eagles as well as osprey and other raptors. Grouse and Mirriam turkeys are everywhere. The waters abound with fish of several species including both large and smallmouth bass as well as the kokanee, a landlocked sockeye salmon. There are trout resident here as well.

It is certainly not difficult to understand why people are drawn here but being as remote as it is from major population centers, and due to its vast size, traffic is amazingly light and other people are rarely seen, let alone allowed to disturb one's quiet relaxation.

This Saturday began as many others had. We had decided to take the boat for a run down the lake a ways to a private little place we liked on the western shore. Being on the western shore assured us of privacy from the landward side as it was land belonging to the Colville Tribes and was off limits to general incursion by tourists. The shore was particularly rugged in this area and we were hoping to have the opportunity of locating some of the plentiful California Bighorn Sheep in the area. Enjoying a calm day, we fished in a desultory manner, catching several of the very delicious Kokanee Salmon that were plentiful here as well. As the sun drifted inexorably west in a sky of perfect blue, we were at peace with the world when I got the idea to play Ron Morehead's "Sierra Sounds" audio CDs with the idea of seeing if we could get a response from the craggy mountains that bordered the lake.

As we drifted along, catching a fish now and again and listening to the strange vocalizations being played at a pretty high volume so as to effect a broad coverage, we were suddenly startled to hear an answer come from far up the mountainside. At first, I thought it might have been an echo because it was so faint, but there! It came again.

It was two diligent people who then attempted to locate the individual responsible for those sounds coming back to us from the mountain. Using my binoculars, I spotted him and watched as he moved slowly down the rocky bluff. Slowly he came on to the sounds. I can only imagine what he was thinking or feeling as he heard the vocal utterances from his type coming from the lake shore. Was he thinking this was an interloper, come unbidden to his land? Perhaps he held the hope of finding a new mate? We have no way of knowing the why of what he was doing, but that he was doing it was now beyond doubt.

Anxiously, we watched as he climbed down the steep banks until he disappeared… we knew he was moving on pretty rapidly and we had timed his movements, so, we thought we would know about when he should reappear down the slope further. It did not happen. He had disappeared and we had lost our chance. It was highly disappointing and a relief all at the same time. I was anxious to see him again while my companion was less convinced. Finally, I determined that if he would not descend further, I would go to him. We could see very clearly all around where he would have to pass in order to egress the area we had last seen him and nothing had passed in our vision. I was sure he had to be there in that low area somewhere and I was just as determined to see him again as he obviously was determined to remain in hiding.

We moved the boat in the cove to about a hundred yards from the rocky shore and I dove in and started swimming to the beach. While I was swimming contentedly along, unbeknownst to me, the big fellow reappeared to my companion, still moving down the rock bluff towards the water's edge. Although she tried her best to shout at me and let me know what was happening, I could not hear her while swimming and continued along at a good clip... down the hill the creature came and inexorably to shore I came...

As I reached shore, I stood up... and looked a very large, very black, very powerful being directly in the eye at a range so short I had no real wish to be here. My guest solved the problem as he looked at me

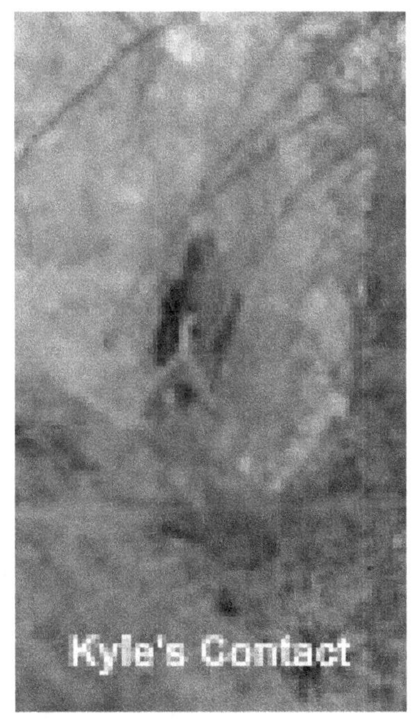

by screaming a very loud, very blood-curdling scream and I was frozen to my spot. I could not have possibly moved a muscle. I don't know if he had immobilized me or if I was simply in a trance, but I know I could not have moved a muscle. I do believe if someone had shot me, I could not have fallen down, I was that rigid. While I was in this state, my friend retreated back up the hill and out of my range of vision and I was released from my paralysis. I immediately jumped back into the water and swam my fastest back to the boat. Those reports from my partner that stated I was up and planing are an exaggeration, however! Fast I was... but not that fast.

Fortunately, my companion was able to get a picture of the fellow, and even though he was far away and the image had to be enlarged greatly, he still shows very well in it.

The BBC Presents

In 1998, BBC presented a program designed to debunk the Patterson-Gimlin Film… in that effort, they had constructed a suit that was the state of the art for that time… the red figure in the photo is that suit… as one can readily see, the arms prove it to be a suit… and, add to that, the head sits up on a pedestal allowing it to rotate on that pivot to look behind, whereas the Patterson figure has no such pivot and must open her shoulder to see behind her… Once again, as

all such efforts have done, they PROVE the authenticity of the film rather than to debunk it.

In this regard the American version of this X-Creatures episode was quite different from the version broadcast on British TV.

The American version (shown on the Discovery Channel) focused attention on the Hollywood costume designer's attempt to build a "matching" costume. The British version left this part out completely. During the costume building sequence, the designer hints at the difficulty they had matching their costume to the figure in the footage. Among other things, they could not match the fur with artificial fur, and they could not recreate the large moving muscles in the Patterson figure. Nevertheless, Packham touted their creation as an "exact match" of the Patterson creature, and even showed the "matching" costumes side by side.

This comparison will be discussed in greater detail and a "proof" of the fallacy that lies herein in a later chapter.

It should be noted that after airing of this "Documentary", BBC stated that it did not, in any way, disprove the validity of the PGF.

I have a friend in British Columbia, Canada who is full-blooded Lakota Sioux. I met him through my research partner in B.C., Brian Bland. Brian

met Randy when he was featured in a documentary concerning the sasquatch people and Brian realized they lived in the same town, causing him to contact Randy and the two began to prowl the wild places together.

When I began exploring Brian's area, leading up to the five-year study we conducted on the sasquatch written language, glyphs, Randy was there to help us when needed. I was quite amazed at Randy's level of understanding of these large beings and ascribed it to his rearing in the Lakota culture. He and I spent many hours in discussion of these fascinating natives.

This incident happened a very few years before I had met Randy, but it occurred in an area I know well. On this particular day, Randy and his son Ray were in the area on a day hike just to enjoy the countryside and get into nature for a time.

The forestland in this area of B.C. has very little underbrush due to the fact that the last glaciation which ended about 12,000 years ago stripped all of the topsoil here and deposited it further south in what is now Washington state where those glaciers terminated. The ground is basically rock, with only a light covering of detritus from the trees that grow there, covered by a layer of bright green moss. Without the brushes so prevalent in most coastal timber

biomes, it makes objects there much easier to see… and that is what happened this day.

 As Ray and Randy walked across this mossy tract under the trees, suddenly, Ray pointed and said, "Look there!" as a hairy figure ran across the brush-free ground. He ran behind a huge cedar stump left behind from the logging done in the 1930's and then turned, peered over the top of the stump, spread the huckleberry vines growing there and looked directly at the two humans watching him.

Randy was able to snap the photo shown here, even though his camera began to malfunction.

Subsequently, we placed Brian in the same spot and photographed him peeking over that stump and found that he, at 6'4" had to stand on a root projection to see over the stump.

On Researchers, Experiencers and Experts

I do not know how many times I have heard the old saw, "In sasquatch research, there are NO experts!" Let me say this to that... They're WRONG! In this section, we will meet many who go far beyond the simple title of researcher or investigator and all the way to experiencer or even expert!

Yes, there are experts in this field... many of them. Perhaps there are none across the entire spectrum of all of sasquatchdom, but there definitely are experts. For instance, in all things anatomical and regarding feet and foot prints in particular, Dr. Jeff Meldrum is an expert in his field. As to how they live their day to day lives and how they structure their families, Arla Williams and I would qualify. For the structures they construct, I would look to LeAnn Carnegie for expert advice. For their written language, glyphs, Brian Bland and myself began the research in that field. For definition of sounds and their breakdown, I would go to Peck White or Alex Munoz in Georgia... and for their oral or spoken language, there are two names that literally mean this…

Ron Morehead and Scott Nelson have brought new and greater understanding to a little understood phenomenon. Together, they have created something greater than the sum of its parts and, in the process, have opened minds to new possibilities.

Their odyssey began in 1972 in a remote deer hunting camp far off any well-worn trail somewhere between Yosemite and Lake Tahoe in California's Sierra Nevada Mountains.

Just to reach this hidden place required traversing steep mountain trails via horseback then turning off onto a route that was not even showing a trail. At the end of this ordeal there was the most primitive of hunting camps… merely slabs of wood leaned against the trees to provide a modicum of cover against the weather, always suspect in these high-altitude environs.

As one might suspect, there were no neighbors within several miles, at least, and the men that came here were strong, independent and very capable. They were not apt to panic at mere whispers nor would the everyday fauna bother them in the least!

It was while they were in this remote camp that they began to hear strange sounds. At times, it sounded like monkeys fighting and at other times, like samurai chatter. The hunters in the camp made a game out of talking to the owners of those voices, trying to entice them in closer, so they could be better identified... a desire that was to remained unfulfilled during their sojourn in these mountains. They saw enough to realize that they were not men as they, and we, would know them, but they could not precisely identify exactly what they were seeing.

Fortunately, Ron Morehead had brought his tape recorder and he proceeded to record the things they were hearing. He was able to get an extensive record of their utterances and, upon the crew's return to civilization, he took the tapes to a newspaperman he knew who the felt could help him shed some light on the subject.

Although the mysterious sounds had begun earlier, even to the extent that some of the people were scared to return to the camp, the first year Ron recorded was 1972 and it was then he approached Al Berry with the results of this work. Al accompanied the crew in 1973, but nothing transpired that year. In 1974, the big, hairy people were back.

Ron will tell you himself, at this time, he had no experience with the sasquatch beings. He was not skeptical, but neither was he knowledgeable about them. He simply had no opinion. This is why he had asked Al Berry to accompany him. First, he was an experienced recorder and, second, he was not someone who believed in or could be swayed by simple tales of a large bipedal primate possibly living in the high mountains of California.

When copious feet of recordings were made, they were sent to Professor Kerlin at the University of Wyoming who then was able to certify that those tapes were legitimate. They had not been tampered with, adjusted, edited nor manipulated in any way. These recordings can be heard today simply by going on line and calling up the website www.ronmorehead.com.

Several years later, a retired Navy Cryptolinguist with more than thirty years of experience had sat down with his son to help him get started on an essay assignment for school. The young man had expressed an interest in the sasquatch phenomenon and Scott suggested they search the web for some references. In the course of this search, the pair came across Ron's "Sierra Sounds" website and the elder Nelson immediately recognized what he was hearing as a language…

Perhaps it would be appropriate at this point to describe exactly what a U.S. Navy Cryptolinguist is and what he does and, particularly, how that applies to our work here... Basically, these people are communications specialists. They sit and listen to radio traffic from people of concern and determine if it is intelligent communication or merely gibberish. Even if it is a language they do not speak, they must be able to recognize those forms that make language and then pass it on to someone who does speak that language to decipher the actual meanings contained therein.

In Scott's case, he speaks three languages besides his native English. He speaks Farsi (Persian or Iranian), Russian and Spanish... but what he heard on these tapes was none of those... it was a completely foreign language from which he could gain no understanding.

After contacting Ron Morehead, Scott Nelson then undertook to create a phonetic alphabet from that which he heard on the Sierra Sounds tapes. That phonetic alphabet is still available today by simply contacting Scott via email, on facebook.com or through Ron Morehead at his website.

I have heard these gentlemen speak on this subject and was so impressed, I invited them to present at my 2012 "International Society for Primal People" conference in Richland, Washington. I gave them two hours to divide as they would and the results were simply magnificent. I have had them return to other conferences, and have spoken opposite them at still more.

At one point in the Richland conference, Scott was called to task by Dr. D. Jeff Meldrum, Professor of Anatomy at Idaho State University, Pocatello, Idaho when he made the statement that there were two individuals conversing on the tape… "How do you know there were two individuals?" he interrupted Scott in his presentation to ask.

"Because," Scott responded, "one did the same thing you just did to me… began speaking before the other was finished…"

I'm not sure how many in the audience caught that slight double entendre, but I did, and filed it away… having not forgotten it even ten years later. There is a video of that 2012 presentation on my www.Youtube.com channel at https://www.youtube.com/watch?v=vaZxJg3GMnE&t=281s

As one may well imagine, these men are experts in what they do. Their work is scientific, it's repeatable and it's reproducible… and those facts make this very potent evidence, indeed.

Barb 'n Gabby

Something over eight years ago, back in 2014, a new research team came to my attention. The first thing I noticed was that one of the team members was a real dog... the senior member, however seemed level headed and very interested, if not too well informed. I began following their work loosely and noticed a distinct improvement in the material being presented in their videos they released regularly... most from near their western Washington home.

It was November of 2014 that I first met this lady at the

Sasquatch Summit in Ocean Shores, Washington where we were both on the dais as speakers for the event. Although we had talked prior to our meeting here, we met, face to face here at the Friday night session. It was but the work of moments to understand that here was someone who was devoted to her work and was interested in gaining more

knowledge for the RIGHT reasons... not for greed or self-aggrandizement, but simply for knowledge. I could see early on that she was chosen of them to

spread their word through her work and effort. During the round table forum at the end of the Summit, I credited her with the saving of my life when an individual was attempting to usurp the proceedings to advance his own

position... every time he said something I disagreed with, I'd reach for the microphone and she'd take it away from me and slap my hands... thereby keeping me from saying the wrong thing in front of this large gathering!

In 2015, Barb, teamed with her partner, Gabby, and fellow species mates, Sandy Nelson and Samantha Ritchie, now passed on, began hosting outings in their research area.

They invited people to join them for a weekend of investigation and research. In June of 2015, I was pleased to be invited to one of these events and found it to be a wonderfully spent weekend in the forests of the Gifford Pinchot National Forest just north of Mt. Rainier. Sandy was a wonderful camp coordinator as well as the chief cook and bottle washer! Sam had her own you tube channel and page on Facebook called "**Planet Sasquatch**".

Sam was a professional at film analysis and had the knowledge and tools to do the work required to bring the hidden nuances of a film to the forefront and to understanding. Actually, some of the work she has done is nothing short of amazing and the results quite spectacular.

Our outing was held immediately north of Mt. Rainier and the few days spent there were very pleasant and altogether too short. Many pieces of evidence were found during our time there and some prints were cast as well. The dinners and breakfasts were all a person could ever want

so time spent there was not to be bested anywhere!

I spent further time with this group in part on our gathering at Wind River over the Labor Day weekend... from one of the most delicious omelets I've ever eaten to a serenade by owls... from beautiful stones demanding to be carried home to tracks cast and rocks thrown into our camp, these few days was all it could be and a great respect was gained by many people making the acquaintance of this team for the first time...

I had planned to attend the last Barb 'n Gabby campout of 2015 in mid-September but a concatenation of ill-fitting circumstances demanded I not be in attendance... and I am so sorry I missed out... The events of this few days spent in the vicinity of Mt. Rainier on the western slope of the Cascade Mountains rivaled any similar period I've ever been associated with at any time. From suspected sightings on first arrival to the discovery of unique and elaborate structures left for them... from track imprints of the tiniest of their citizens to the most honored of their monuments... Many were the marvels of this outing.

One of my favorite events occurred shortly after the arrival of the Barb 'n Gabby crew to the area when a couple who had been camping for several weeks in the area came to them to warn them of the "Unusual Bear Activity" they had been experiencing. According to the accounts given by this couple, a very large grizzly bear had been walking around their camp at night, disturbing their camp gear, rustling and looking into their tent and leaving tracks so large they could, themselves, stand in them. The first problem with this scenario is that there are no grizzlies in this part of Washington. They occur only along the northern tier, adjacent to Canada and in the far northeastern corner, near Idaho. Second, these tracks had no claw marks and grizzlies have non retractable claws... they show in their prints. Further, if a grizzly came into camp, it would be in search of food and the destruction would have been widespread... possibly even injury to the occupants. Grizzlies are not good campmates and the best course of action one can take if they do show themselves is retreat... as quickly and thoroughly as possible, get out of his way. In all the land he inhabits, Mr. Grizz is the top of the food chain and has been for millennia... one should never challenge him under any circumstances... simply retreat and stay out of his way. There is nothing in our world that can turn a simple stroll to the outhouse into the adventure of a lifetime like one well-placed grizzly on the path!

Barb and others of her group undertook the task of explaining the actuality of the tracks they had found around their tent to the couple.

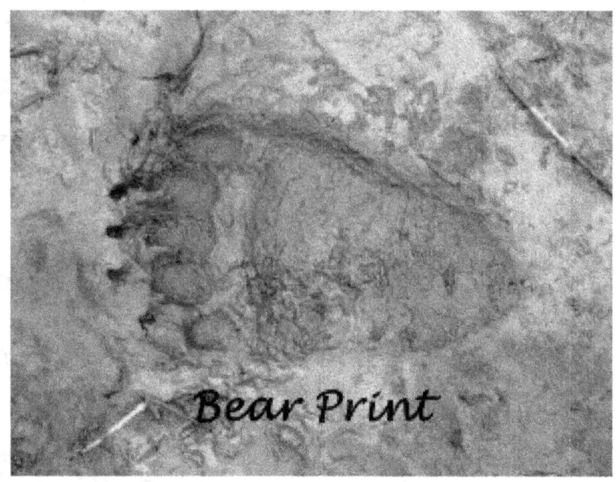

They were shown the shape of the tracks found and how that would differ from the track of a bear. Even a double struck bear track, one in which the bear's hind foot lands in the impression of the front foot as the creature moves on forward would not have accounted for the extreme size and square shape of these prints. Tracks were found in the immediate vicinity that exceeded fifteen inches in length. Applying the standard anatomical conversion factor for primates, this would yield an individual in excess of eight feet in height that would weigh between seven hundred seventy-five and eight hundred pounds!

It should be noted that a bear's rear foot is triangular in shape as per the illustration. It will nearly always show the claw marks... especially a grizz because their claws are not retractable. Further, a double struck track is seldom even. There is usually an overstep to one side or the other leaving it less than uniform in appearance. A sasquatch print will be quite uniform in width from toes to heel with the front half, that portion forward of the mid tarsal break, slightly deeper

in the soil than is the rear portion. A close examination of the mechanics of their gait will show why this is.

The sasquatch does not have a stride like a human. When we walk, we pole vault over our hips to land, heel first, rather heavily. Our heel strike is our deepest penetration and the ground penetration lessens as we roll forward to the ball of our foot and onto our toes for another pole vault with the next stride. Our feet strike the ground at roughly our shoulder width or a bit closer. The sasquatch uses a compliant gait in which his foot is lifted from the ground, swing out to the side and brought forward to be placed down smoothly and in line with the track made by the opposite side foot, with the entire foot making contact with the ground simultaneously, placing about seven and a half pounds per square inch pressure on the soil. As he moves forward, the joint that is the mid tarsal break allows the rear half of the foot to rise clear of the soil and transferring the weight held by the entire foot to the front half of the foot, increasing the ground pressure significantly, perhaps to as much as fifteen pounds per square inch, creating a deeper impression in the front half. This is something to always look for in a casting to determine at a glance if the track could have been man made and left there to be found.

In an area very close to where John and Abby had their tent erected, was found a most curious structure. It closely resembles a fence section or a gate and it spans a small stream there. According to John's personal testimony, the structure was not in place the evening before the "unusual bear activity" and the appearance of the oversized tracks that surrounded their tent. When they looked the morning following this incident, it was, in fact, there as well as a tepee type structure that had since been removed. If one watches Barb's video closely, he can hear John state clearly these facts. He further states that no one else was in this area other than he and his wife until Barb arrived with her entourage just prior to their joining the Barb 'n Gabby group in their camp. Also in great profusion in the area are tree arches with the top ends being held down by as many as four separate logs. These appear to be 

live, viable trees that have been bent to the ground and pinned there...

Gabby is gone now, and missed greatly… she passed a few years ago and has, eventually, been replaced by Goldie. Barb 'n Goldie have a wonderful thing going here... I highly recommend her work as something to follow and emulate. Are all of her conclusions 100% correct? I doubt it seriously, but then, I doubt mine are either! But... she is a credible gatherer of data... she, with her team, do some marvelous things with the video evidence. Do I expect their work to provide definitive proof of the sasquatch people to the public at large? I pray that NOT happen ever... but for the serious investigator and the casual observer, her work provides another avenue of good data that has been well thought out and analyzed in a logical and non-emotion manner. If anyone is NOT a subscriber to Barb's Youtube Channel, **"Squatchin' With Barb 'n Goldie"**... I highly recommend you become one soonest... I have!

The Howler
Richard Stewart Taylor

It is very difficult for some people to believe or accept that there are a substantial number of folks spread across this nation who have regular interaction with the Big Guys of the forest. This was not garnered by brief excursions into the woods yelling, hooting, and banging on trees with the expectation of these subjects appearing and performing like the poor carnival organ grinder monkey tied to a chain and collecting coins in a cup for his master. This common mindset will often lead to absolutely nothing but is used time and again without lasting success.

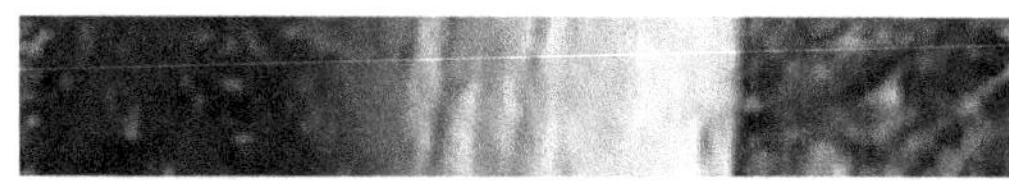

I don't claim to have all the answers, but one thing I do know drawn from empirical knowledge, these subjects are not the dumb animals some believe them to be, yet these same folks have NEVER even seen one! I do not argue that point, people will believe what they choose to believe and that is their choice to make, not mine.

With that said, I was visiting a full-blood Cherokee tribal member in a remote forested area south of Jay, Oklahoma back in 2016 with two other experienced researchers. The small, local community was familiar with the local Big Guy they referred to as the "Howler". They knew he was there and left him alone, but occasionally had a calf, a chicken or a dog messed with or taken. The name "Howler" was given to this subject because they made howling calls and noises at times when they passed through the area. I was fortunate enough to record several while I was there.

As has been the custom of some Cherokee people going back thousands of years, this landowner would place edible gifts along the perimeter of his fence line for the Howler. In turn, the Howler left his dogs, chickens, and livestock alone. I found one of these jars of peanut butter sitting on a fence post along his fence line. I mentioned to him the lid was still on the jar. He said the Big Guy places the lid back on, so the varmints won't take it.

I opened the jar and found about two-thirds of the peanut butter gone with a distinct thumb push and index finger swirl in it. I asked the landowner to hold the jar while I shot a picture of it as a comparison to the man's hand and digits and have attached photos of this jar.

I also took two videos. One was taken along a remote ridge as I heard the Howler coming up through the adjacent, extremely steep ravine. There was no way to safely scale down into it, which is why the subject used it. They kept making this "Whoop-you who" call. Now for those who might say, "How do you know it was the Big Guy?" The video recording does not do justice to the resonance and strength of the whoops. One could feel the sound shake the woods and my body even at that substantial distance. Nobody I know has that kind of lung capacity except the Big Guys, unless they used a sound system with four JBL woofers!

We were also observed later that night by a large subject hiding along a ridge to the property. It was crouched down… maybe on all fours… in a thick patch of foliage watching us. I knew exactly where they were and that's all I care about. Even during the cover of night, these subjects maintain cover and concealment.

Bottoms Down

It was September of 2018 and it had been raining so steadily for so many days that people were beginning to discuss cubits and wonder about the accessibility of Gopher Wood… the animals were beginning to pair off in order to come from the hills two by two… it was very, very wet and the gang from the "Squatchin' with Barb and Gabby" group who

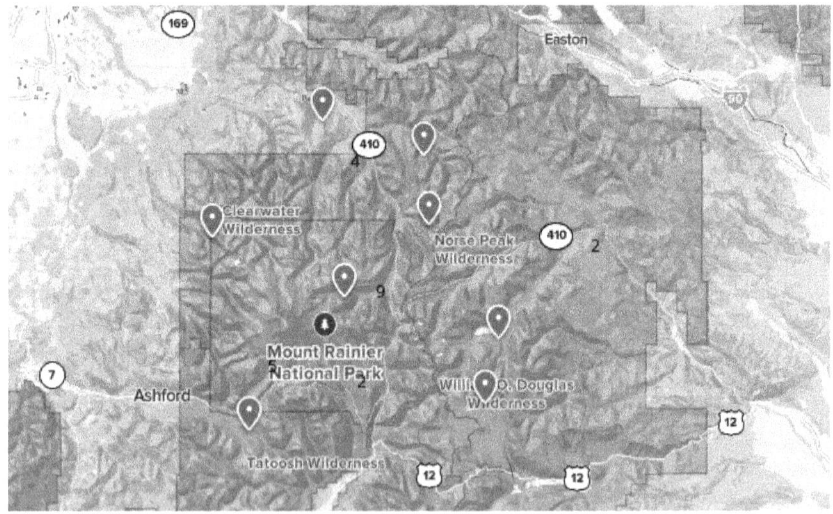

were in camp in the middle of Washington's Cascade Mountains were getting a bit waterlogged.

On the fifth day the skies lifted a bit and the gang decided it'd be a nice time to see what was available to find on the

beach of the lake situated in the ancient magma chamber of a long, long dead volcano… not one of the active volcanoes found here like Mt. Rainier, Mt. Saint Helens, or Mt. Adams, the three that effectively surround most of the southern Cascades. This lake is an irrigation impoundment on the Yakima Valley Irrigation System so, by this time of year, the water level is quite low. In fact, it is so low that it is possible to walk completely around the eastern side of it on the very wide, flat beaches. There will be creeks to cross on the way, but nothing that is impossible, improbable or even difficult.

As our crew spread out to see what clues were left in the super soupy mud of the bare beaches, some moved further away from that area most frequented by guests. Two of these, Kevin Carney and Ashleah Stinnett found themselves removed from the others when Kevin decided to climb a larger than normal stump to see what he could of their surrounding area. When he had done so, he looked down to see the most improbable of scenes laid out before him... there on the beach was what appeared to be a pot of pure mud and next to it, a most perfect impression of the posterior of a human-like being...

As can be seen in the adjoining photograph, whatever made this scene had sat, naked, in the mud and dangled its feet into the mudpot. When it arose, it rolled to the left in this picture and placed its knees on the ground just behind the butt print. In doing so, it placed its hands on the ground to leverage itself to a standing position. All of these conclusions are readily reached by simply looking at the imprints on the ground.

One of the first things noticed was that the mudpot was artificially filled with water by more than one of the individuals scooping water back up from the small ditch of run-off water that was very close to this structure.

Next, when one examines the "seat" closely, some things become abundantly clear... first, it was a female because the labia are quite evident as is the anus. Second, the individual was quite hirsute. The adjacent photograph shows the hair striations quite clearly below the left end of the tape measure all the way to the bottom left of the picture. They are also evident to the right, directly beneath the "10" mark on the tape. Also easily found in the print are the labia and the anus. What is significantly missing is any indication of the coccyx.

In total, three individual butt prints were found. The one pictured here is of a female. A smaller male was found and cast as well and a third that could not be defined was also cast, but not cast. In the cast of the young male, the scrotum is quite visible as is the anus. In none of the three was the coccyx seen. The anus was visible in all three as were the hair striations.

Subsequently, we enlisted the services of one of the group members and had her emulate what these individuals had done and a casting was made of her print as well as that of her husband. In

both of those, the body parts mentioned, the labia on hers and the scrotum on his as well as the anus was visible in the cast… the coccyx was also very prominent in both of the human casts. There were no hair striations evident in the human casts.

Casts were also made of several non-human footprints in the area as well as some hand prints. Unfortunately, probably the most valuable hand print was lost when it was stepped on by an over-zealous onlooker. It was at this point that we began rigorous scene control of our investigation scenes using police forensic tactics. We began marking off the affected area and using ribbon to delineate the "no entry" zones to preclude a recurrence of this event.

Chapter 7

The Final "Proof"

We have, heretofore, discussed a multitude of encounters and cases where stories and cases were corroborated by more than one person observing the being. We have met some very fine investigators who know to conduct said investigation and we have met investigators from outside the normal scope of things. Is it possible ALL of these people are lying for some unknown, perverse reason? It kind of boggles the mind to think that this could even be possible, does it not?

These people cited understand that, even though it may be theoretically possible, that does not mean that is functionally practical to do so. This is a most important concept to understand and accept! Just because is doable in the lab does not mean it is something that can be done in the field! Further, even if it could be done, is it wise to do it?

Although we peeked at the concept of Intermembral Index (IMI) in Chapter 3, we are now going to examine it in detail and show examples of it in use or various species and various conditions. In this treatise, I will be referring to the work being done currently by the analyst, "Thinker Thunker" (TT) and showing some of his work in this respect.

TT has a rather extensive Youtube.com channel that features a multitude of his work. The fellow is an artist who states he was initially drawn to the IMI concept because of his knowledge of human body ratios and instantly recognized the difference in those ratios in bears, chimpanzees, gorillas and sasquatch.

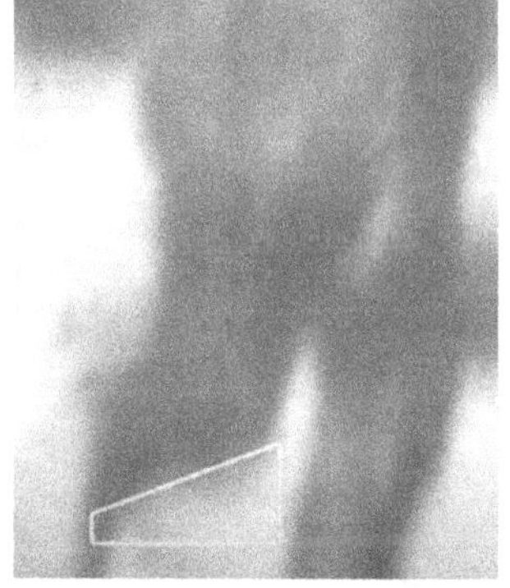

The first anomaly he addressed quite some time ago in his earliest videos was the angle to which we lift our leg when walking, vs. the angle their walk demonstrates. According to his data and demonstrations, when a human walks, we lift our leg to an angle approximating fifty-two degrees (see figure above). When the sasquatch person walks, he lifts his legs to an angle of seventy-three degrees (see figure at the side) … a difference of twenty-one degrees!

While these data were interesting and certainly helped in many cases, it was still safer to rely on the Intermembral Index when and where it could be obtained.

Thinker Thunker has even added to this concept by adding in the length of the torso, and, I have to agree, there are times when that helps in our identity. TT has named his data set "Proportion DNA" or, simply, PDNA. I have been using his methods and adding them to my own and I have found a very reliable method for determining the validity of photographs and videos of purported sasquatch people.

To illustrate TT's methodology, let us examine the adjacent frame from the Harley

Hoffman video. This is a random frame from which a screenshot has been taken for purposes of research.

The

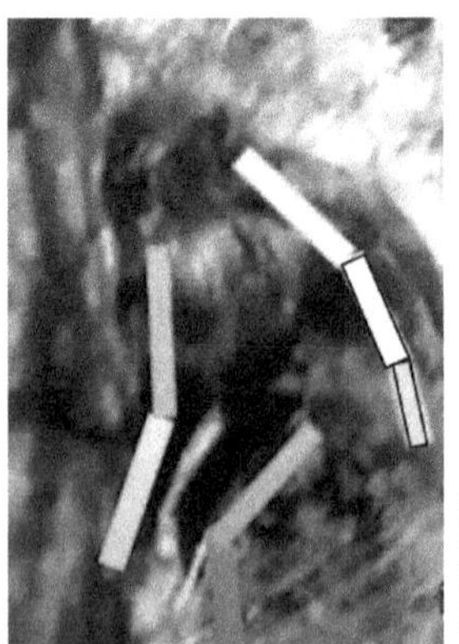

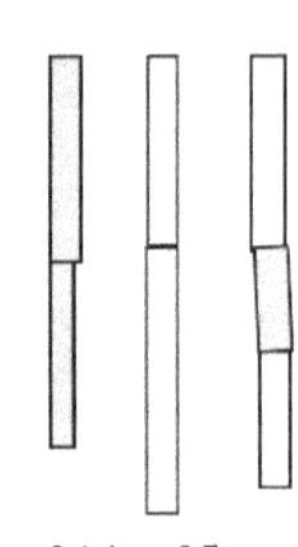

second photo frame shows this same figure with lines drawn on the arm, from point of shoulder to wrist, from Atlas vertebrae at top of spine to tailbone and then from hip to ankle. These individual lines are then straightened and aligned showing a line each for the leg, the torso and the arm.

The illustration to the right now describes the ratios of the arm length to the leg length to the torso. In comparing the arm length to the leg length, we find we have an arm 3.1 units in length and a leg 3.7 units of length. It should be noted here that the actual units are immaterial… they could be centimeters or miles and the resulting ratio would be unaffected. At this point, we can divide the 3.1 units of arm by the 3.7 units of leg we arrive at a ratio of .84… out subject's arm is .84 of our subject's leg. To get our IMI from this, we simply multiply by 100 to clear the decimal, hence IMI is .84X100=84… our IMI is 84 for this being and that is decidedly not a human… who would have an IMI in the 72 range!

Understand, please, that we are plus or minus 3% accuracy here. Also, we do have to make estimates where our subject joints actually are, but 84 is not, nor is it even close to 72 within any degree of allowable error!

I have illustrated the torso measurement here because that is how TT did it in his analysis and we will discuss a case where that became important at a later point in this missive. It was due to TT's work with this that I realized the full import of my own work and realized the degree to which this bears importance.

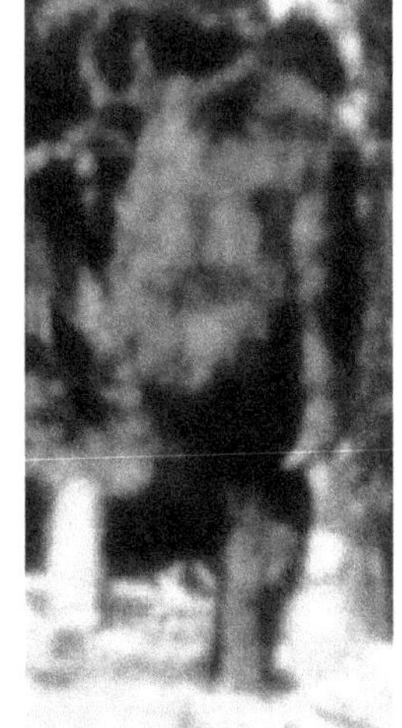

It was at this point that I decided to apply these principles to some other famous and near famous films and photos. Of course, the initial one had to be my friend Bob Gimlin's film from 1967. I find there are few people who have not seen that film and do not know the name of the Patterson-Gimlin film. For this event, I decided to use Frame 72 of that film since it had a good view of both the arm and the leg with no curves nor bending. In all, they do not get much better to analyze than this one!

Patty, being the willing subject she was, yielded an arm length of 2.0 units and a leg measurement of 2.376 units. When we divide 2.0 by 2.376 we get a resultant of .84 and then multiply that, as before, by 100 to clear the decimal, we have an IMI = 84... again, definitely NON human once again.

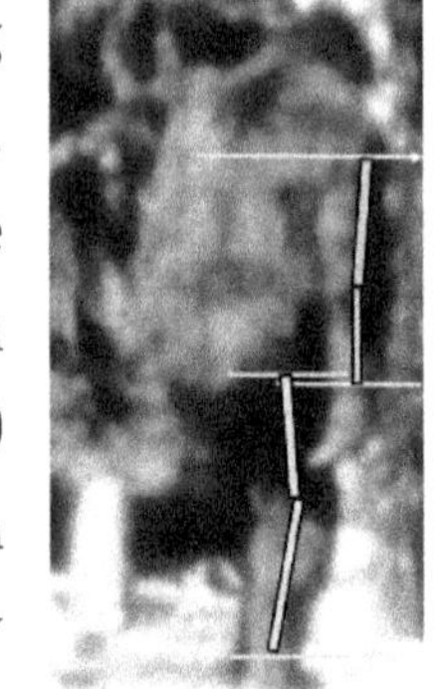

Several years ago, BBC in the UK advertised the ultimate debunking of the PGF... They had a crew come to them for the money to commission a suit to be made that would make people understand, finally, just what a hoax had been perpetrated by these two cowboys who had generated this film. They never stopped to consider that they were about to use technology that was not available in 1967, nor for many

years thereafter… at least they didn't allow that fact to alter their stance on the matter to any noticeable degree.

After spending tens of thousands of dollars on this state-of-the-art suit, here is the result! While this photo is not too clear, one of the first things we notice is that the individual in the suit is looking back at the camera WITHOUT opening his shoulder… his head is on a pedestal… the neck.

In the side-by-side pictures, other factors become evident. First, Patty's arms are substantially longer than the guy in

the suit. Secondly, she cannot look back without opening her shoulder because her chin hits her shoulder, effectively stopping her swing.

While these effects are abundantly clear to any who would look with an unjaundiced eye, there are many who are not so endowed, it would seem, for the hue and cry was loud that the "proof" had been aired that the PGF was fake… not for anything the program showed, for even BBC admitted that the program did nothing to discount any evidence presented in the film. No, the cry came simply from those who had heard the advertisement for the coming show… "PGF Proven False" and accepted that without question. "Oh," this group said, "It's what we knew all along… it's about time it was put to bed anyway!

Now, let's take this one step further and prove where the fallacy lies…

Again, the person in the BBC suit demonstrates the classic IMI of 72 as demonstrated by all humans. While the large, dark and hairy one exhibits 84 we would expect from her kind.

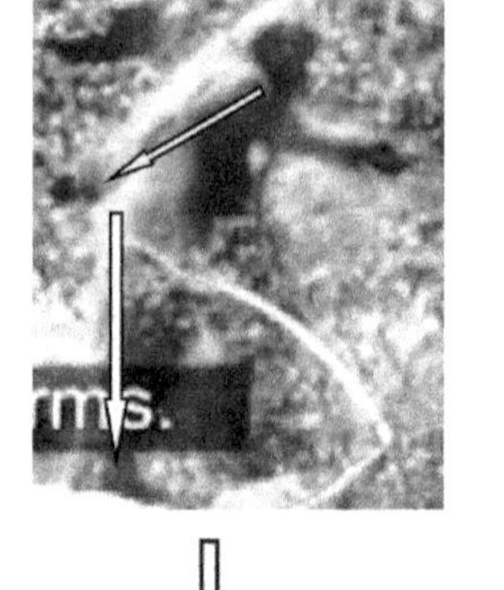

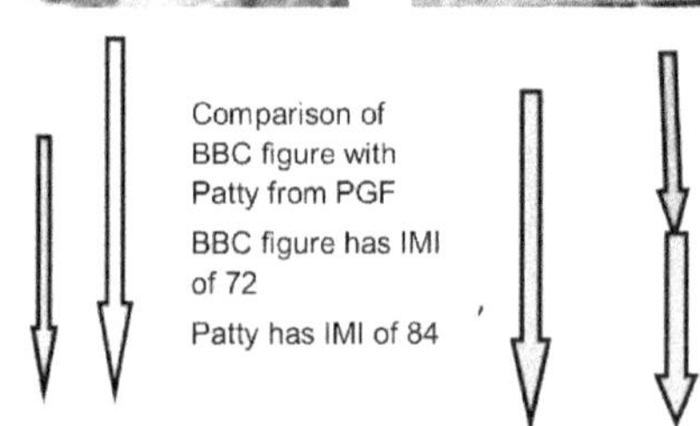

Before we leave the PGF and those involved therein, there is one more "character" to examine… Bob Heronimous has stated that HE was the person in that suit and Phillip Morris has stated that it was one of his suits they used. I have a photo of that suit with

Mr. Heronimous in it, let's test it, shall we?

We can see here that, aside from that suit having more folds than a fat man's chin, it is preposterous to think this is anything more than a man in a suit… but, since we are dealing with evidence allowable in a court of law, may I draw your attention to the head, especially that region just above the brow ridge. If we recall our discussion from a prior chapter dealing with Bill Munns' testimony, in Patty, the head sloped markedly back from the brow ridge, not vertical as does this one.

Next, the step length here, common for a six-foot-tall man, is approximately thirty inches… not the FORTY-EIGHT PLUS inches as measured by four independent investigators at the sight of the filming in 1967. What would this look like if he were to stretch another eighteen inches from this point?

Lastly, our proof… when measured as in the prior photos, the IMI of this individual is 72… not the 84 measured in the Patty figure he was saying he had created.

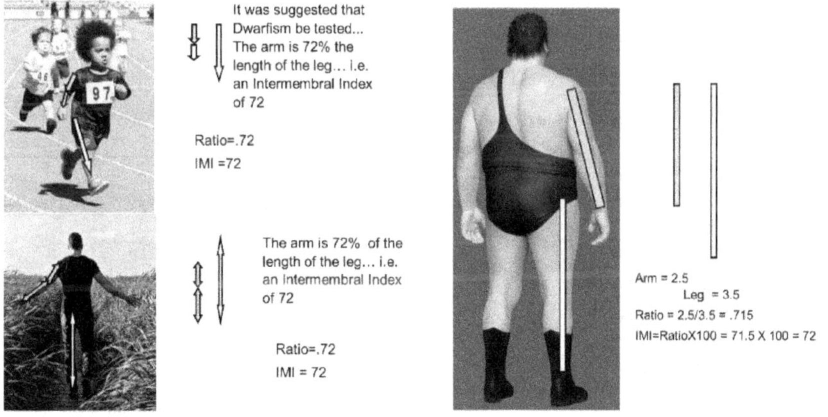

The rest of his testimony is of a similar vein and is completely beyond the realm of believability. It is available publicly should anyone wish to study it further.

"Well now," I was asked… "if you can do this well here with normal people, what about someone with gigantism?"

It was an interesting question, I thought, so I searched out a photo of someone so afflicted… What I came up with was the famous French Professional Wrestler, Andre the Giant. At seven feet, four inches, he certainly qualifies in the size department… and he is afflicted with the gigantism condition. As can be seen here, when applied to his huge frame, the IMI is the same as any other human I have ever tested.

Okay, Thom, that covers that condition, but what about dwarfism? Will it be true there?

Again, I searched out a photograph I could use as a test subject and here is what I came up with… as we can see, the IMI in a person with dwarfism is no different from he who is normal or who is afflicted with gigantism… it's 72!

Next came the questions of Aliens… and, since alien photos are not as readily available as are the other forms of humanity, I had to search… and I was successful, I believe… although they seem to have their own IMI value. Actually, the only thing I could swear to in a court of law is that they are not human.

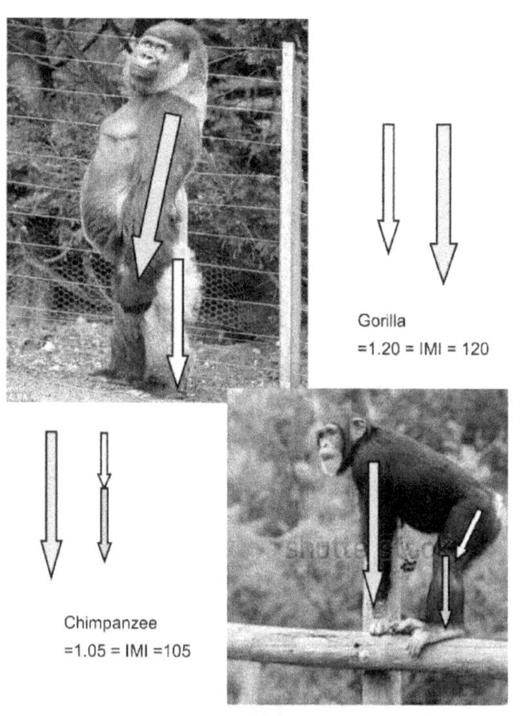

At this point, my curiosity was aroused as to other animals, so started investigating some of those, especially other primates. The first I did were Chimpanzees and Gorillas. I found that each of these species had their own IMI, so, since bears are sometimes included in this mix, I tested one of those creatures.

It is immediately evident that a bear's "arms", or front legs, are equal in length to their hind legs, thereby yielding an IMI of exactly 100. One large difference here is the length of their torso in comparison to their limbs… it is substantially longer than the legs, making some identifications much easier from this simple fact.

Since we now have a reliable way to tell if our subject is a human in a suit, the next idea was to test some of the beings I have been wondering about over the years. These are film subjects that have been debated ad infinitum… even ad absurdium… with no real conclusions reached.

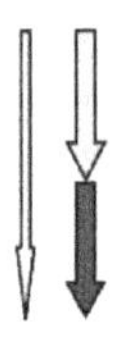

Before starting this exercise, let me explain that most of the photos we see here are screen captures off an ancient video on Youtube.com which are then further manipulated in order to show in this print medium, further reducing the clarity of the photo. Understand, in all of these examples, I was able to identify or closely estimate the point of the shoulder, the wrist, the point of the hip and the ankle clearly enough for this exercise. As of this writing, I still have not been able to do so for the

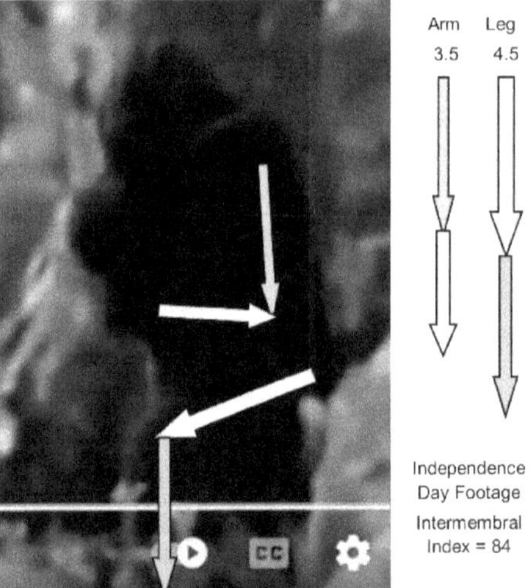

individual in the Freeman Footage nor in the Memorial Day Footage. I have not given up on these, but it will take longer to attain usable photographs in those two cases.

Let us begin with a video readily available on Youtube.com known as "The Independence Day Footage." This is especially appropriate because when I first saw this video, I dismissed it as a hoax. It was not until I had spent a great deal of time studying the individual's gait that I decided she was walking with a compliant gait, not our pole-vaulting gait. At this point, I decided that this could be real, but the proof waited until I was able to isolate a usable frame and apply the test of IMI… my work yielded an arm length of 3.5 units and the leg yielded a length of 4.5 units. Doing the math on these numbers yields an IMI of 84… NOT the IMI of a human!

Being successful with the Independence Day figure, I moved to another that I had reservations about. The Marble Mountain Figure was videoed by a group of Boy Scouts on an outing in Northern California, in the Marble Mountain Wilderness area, roughly between the small towns of Orleans and Etna not far from Oregon.

This video shows that the scout group had found a structure of some kind in the wilderness area then spotted a large, haired figure walking along the ridgetop, gesticulating wildly from its position to the gathered humans. During the course of this video, the figure descends the ridge while remaining perfectly sky-lined on the crest of that ridge.

The calculations yielded an IMI of 82 for this figure, well within our plus or minus 3% error I allowed on this exercise... definitely NOT a human in a costume, is it?

For the next exercise, I shall use some of Thinker Thunker's figure to show that we are, indeed, working on the same basic principle.

Pictured is the Harley Hoffman figure and, again, with arms at 3.1 units of length and legs at 3.7 units of length, the ratio between the two is .84 which we multiply by 100 to yield an IMI of 84... decidedly NOT a human!

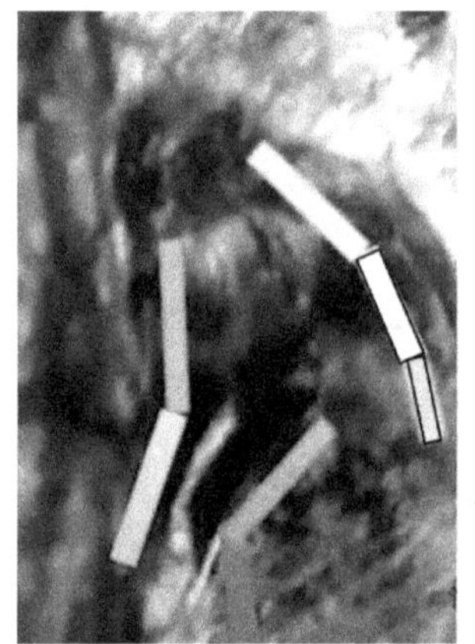
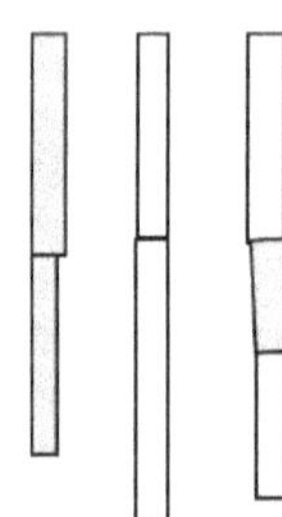

Arm=3.1 Leg=3.7
Ratio=3.1/3.7=.84
IMI=RatioX100=84
Figure is NOT Human

It should be emphasized here that I use the term "units" without specifying what those units actually are. They could be centimeters or they could be miles, as far as we know here. The rationale for this is simple... it does not matter what they are, as long as both arm and leg are measured in the

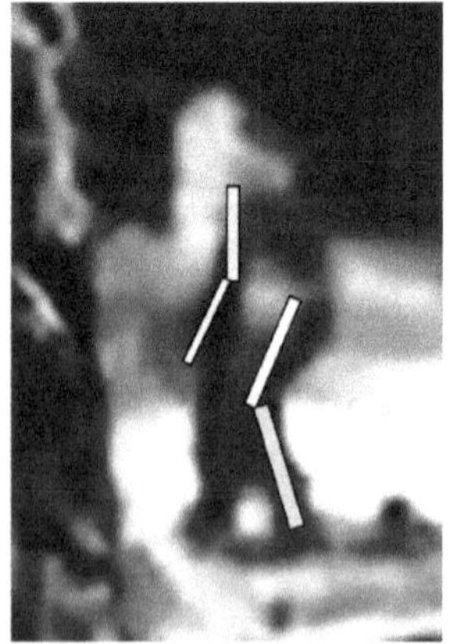
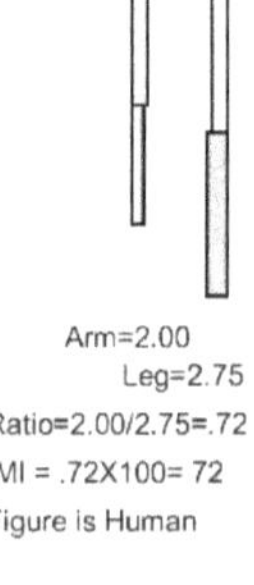

Arm=2.00
Leg=2.75
Ratio=2.00/2.75=.72
IMI = .72X100= 72
Figure is Human

same unit. The actual length has no meaning in this exercise as it is only the ratio of one limb to the other that counts!

At this point, I was feeling pretty secure in my methodology and decided to go up against some I was less sure of, but still had questions about…

There is, in the research world, a group who makes and or distributes film and this is one of their offerings. I've never found a legitimate film with their logo on it, but, after, hope springs eternal, does it not? So, without much real hope, I started in on their latest offering… and the result is herein shown… and arm length of 2 units and a leg of 2.75 units. When we run these numbers, we find an IMI = 72… this is a human in a costume… and we are right back where we started once more with this group!

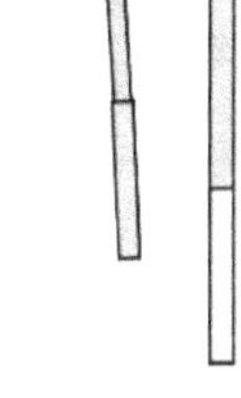

Arm = 2.25
Leg = 3.125
Ratio = 2.25/3.125=.72
IMI=RatioX100 = 72
This figure is Human

Thinking I needed one more success, I took a photo that an acquaintance posted as "legitimate" and tested it in the same way as all the rest. I have spoken with this fellow at several different conferences and, while I did not accept his view of the large, hairy people as being the position of all "First-Nations" people, I've known far too many with alternate opinions to accept his as gospel, I did think his image was worthy of scrutiny…

The image was clear, concise and my critical points were extremely easy to identify. It did not take rocket science to determine that, with an arm of 2.25 and a leg of 3.125, this individual has an IMI of 72, placing it directly in the realm of a human parading in a suit of some sort or another.

 Our final exercise in takes place on an individual that was argued vehemently from both sides and, to be truthful, I could not decide. I saw the viewpoint of those who were saying it was not a giant hairy beast, but merely a bear with mange… the trouble with that is, I tend to be wary of conditions of HEALTH as a factor in determining exactly that which we are seeing.

The Jacobs figure has been around for several years without there being a definitive conclusion, and this was needed. Again, it was Thinker Thunker who did the first examination of this offering and my work mirrored his exactly.

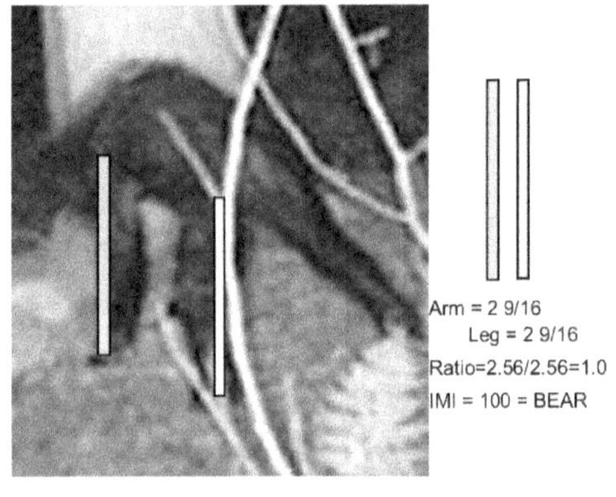

It is immediately obvious here that the front limbs are the same length as the rear limbs, yielding an IMI of 100 and we discussed one creature with that IMI early on in this chapter… The individual is a bear. It should be noted here that TT's use of the Torso measurement… in this case, considerably longer than the limbs enabled him to simply compare the graphs of the three factors and that long torso stood out at once!

As we can see, it is extremely easy to debunk fake photos, which would make me wish those would simply go away… but I doubt they will anytime soon. At any rate, we now have a very powerful tool at our disposal to prove which are humans masquerading and those that are actually large, hairy individuals.

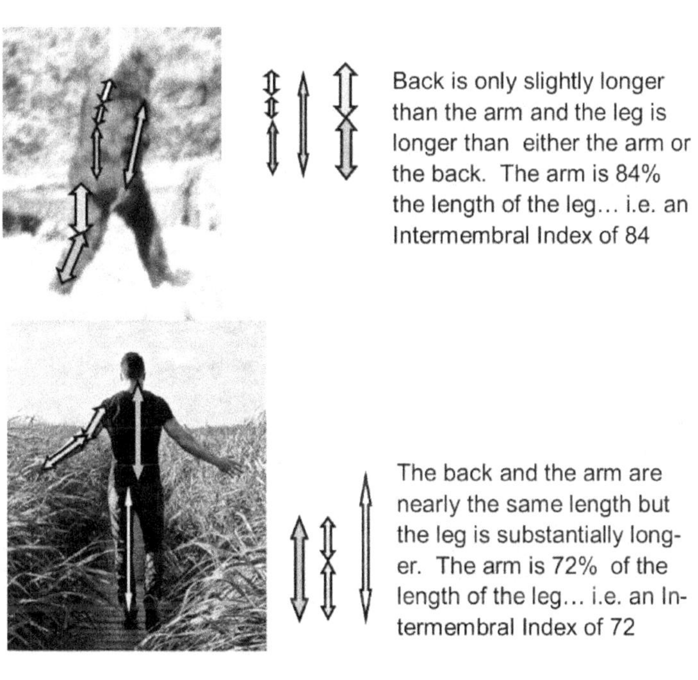

Back is only slightly longer than the arm and the leg is longer than either the arm or the back. The arm is 84% the length of the leg… i.e. an Intermembral Index of 84

The back and the arm are nearly the same length but the leg is substantially longer. The arm is 72% of the length of the leg… i.e. an Intermembral Index of 72

Cry in the Night

In the decade of the 1970's, beginning in 1972, a group of back country deer hunters in the Sierra Nevada Mountains of California experienced an interesting phenomenon when they began encountering large, bipedal creatures around their remote, pack-in hunting camp. One fellow got the idea to record the unique vocalizations they were hearing.

Ron Morehead knew he had something unique here, so he sent a sample of his work to a newspaperman friend of his, Alan Berry. Al was extremely skeptical of what he was hearing and insisted on being included on a future venture into this place. In 1973, this trip came to fruition but there was no repeat of the sounds.

Fortunately, in 1974, the beings were back and for several years thereafter. A

Al Berry & Ron Morehead

sample of these "Sierra Sounds" can be heard on <u>Youtube</u> at <u>https://www.youtube.com/watch?v=lnq4yK1_AOc</u>. Ron even had the foresight to have his tapes vetted by Professor Kerlin at the University of Wyoming against any possibility of tampering or alteration of the tapes.

Some years later, R. Scott Nelson, a retired U.S. Navy Cryptolinguist with over thirty years' experience was asked by his son for help in getting information for an essay for school. The two sat at the computer surfing lightly across the internet looking for some information on sasquatch that might be fitting for the young man to use as a source on the subject.

A Cryptolinguist in the navy is a language expert. His duty will require him to sit in the cans (Navy speak for "listening with earphones on") for long hours listening to transmitted radio traffic. His first order of business is to decide whether what he is hearing is, using the tools and hints of his trade, real language or it's merely gibberish. Even if it is a language he does not speak, he must be able to recognize it in this way. If he determines it is language, it will be copied and shipped to someone who does speak that language.

Interestingly enough, only HUMANS can speak in gibberish… no other animal is able to do so. There are no primates other than man who can even come close to articulated speech. Yes, they can make sounds… screeches, squeals, squawks and hoots, but not articulated speech.

As the pair sat and checked one thing then another, Scott happened on the "Sierra Sounds" website. When he heard what was contained therein, he stopped and said to his son, "That's LANGUAGE they are speaking there!"

He replayed it time and again listening in more depth and with a clearer ear. Each iteration convinced him more of what he was hearing. No, he could neither decipher it nor translate it, but he knew that what he was hearing was not mere gibberish, but an actual language.

Scott then contacted Ron Morehead and received permission to work with the audio recordings, even obtaining 1st generation copies of them to facilitate his work. One of the first things he discovered was that, for the Morpheme streams to begin to appear, he had to slow them by more than 50% of their standard playback speed! Listening again with this outrageous reduction in speed in effect, he began to hear the individual Morpheme streams.

From these beginnings, Scott has developed a phonetic alphabet for their language and continues to teach about and research this subject even as a professor at a midwestern university.

As far back as the mid 1970's, I often heard long, even howls at night. I heard them in Alaska, Canada, Washington, California and even in Georgia and Oklahoma. Early on, I had no idea what they were, so asked anyone I could get to listen to the poor recordings if they knew… I was assured they were, wolf sounds… in lands where wolves had been absent for more than a century… or fox… or coyote… or dogs… or bobcats… or … get the picture here? They were attributed to anything possible besides what actually made them. That took many more years to bring forth as evidence.

When I learned how to look at these howls intrinsically... through spectral analysis, the evidence began to appear. I

was not any kind of an audio expert, but I knew people who could do the work for me like Paul Graves, Peck White and Alex Midnight Walker. The first thing I learned was that these howls did not match ANY other creature on earth! They matched none of the animals mentioned above, nor did the sounds match bears, lions, tigers nor elephants!

The illustrations shown here are, first, the waveform of the unknown BC howl itself, showing

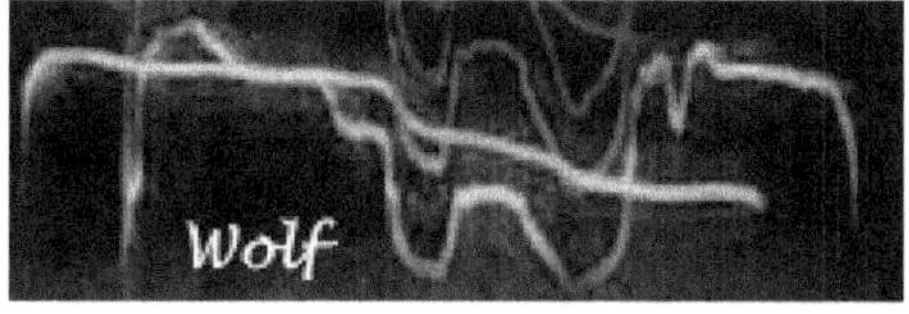

the uniform nature of the sound as described by the spectrogram. Number two shows that of a wolf and the final one is of a human

scream. Notice that the wolf and the human are modulated

with visible peaks and valleys in the sound while one, the BC Howl, is flat, not modulated. In fact, I know of no other source animal on earth where you can receive unmodulated calls from wild things.

While this exercise to this point was interesting and provided superior proof that some species exists that is not classified, it was but one more piece of like data... until the analyst know as Thinker Thunker (TT) got the idea to slow down the replay of the audio…

The result was spectacular! From the original at its normal

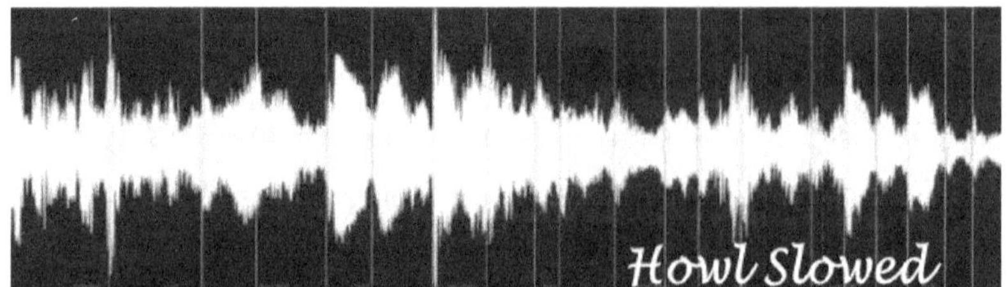

speed to this lower speed reveals, of all things, LANGUAGE! Those howls are NOT merely a steady pulse of tone for an extended period of time, show that within that howl, it is modulated speech! If we compare this to the Sierra Sounds from Ron Morehead and Al Berry, what is different? Now, let's take this one step further and I will share a copy of my own voice taken from one of my Youtube videos I have produced.

As can be readily seen, the only visible difference that can be

detected by a breakdown to its elements are the relative speeds of the utterances… i.e. the speed of the speech. It appears that the howl is a very rapid form of speech… the

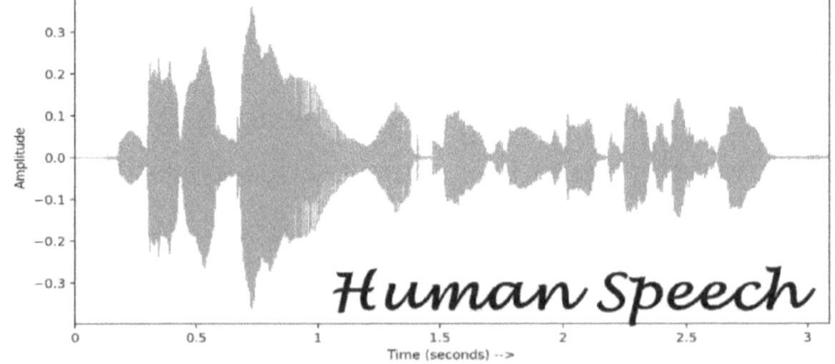

Sierra Sounds a somewhat slower form of speech, and, of course, I know my utterances were the definition of speech… the control factor for this group, so to speak.

Since we can find no known, cataloged animal in all the recordings we have made that can or do produce these wave

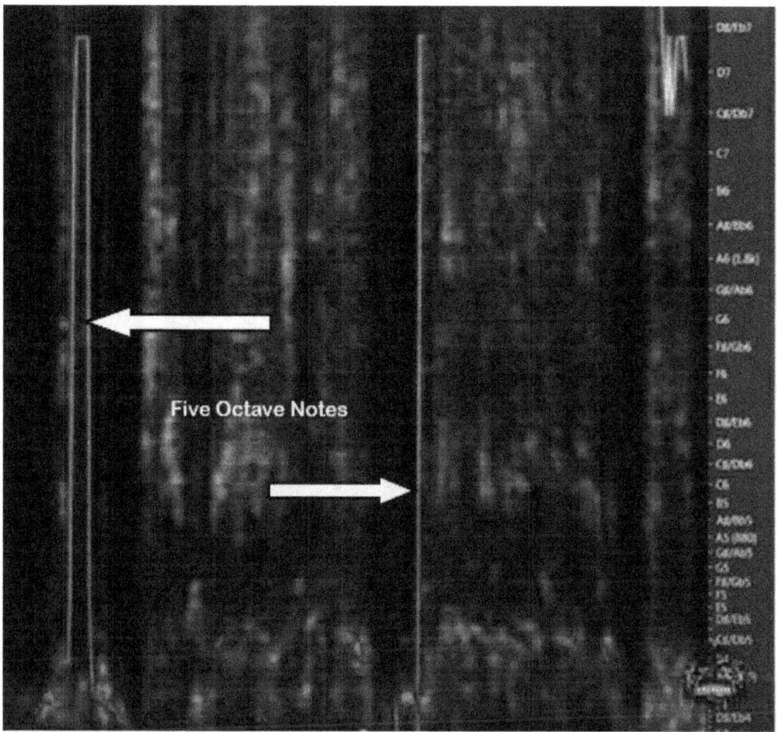

forms it becomes fairly obvious that this is either human or a non-human, unclassified entity.

To continue this line of reasoning, let us look at one of the Sierra Sounds utterances in greater detail. Again, this photo is from a video produced by TT and the credit for its discovery is all his.

The image at the right is a spectrogram showing two instances of a single note spanning more than five octaves! Yes, there are humans whose voices can span five octaves, but even they are rare... Humpback whales and some other animals have vocal ranges that can cover five octaves, but there are NO human voices that can produce a single note spanning those five octaves! Likewise, there are no animals so capable... the utterances from these beings are the ONLY case where a single note can be expressed across so wide of a range.

At this point, it would appear that there is a non-classified species living in our world! To think otherwise in light of this evidence would not seem to be normal thinking, would it?

Chapter 8

Summation

From Testimony to Proof

We now have the testimony... we have heard from expert witnesses on the Patterson-Gimlin Film... and others. We have heard from people who have had encounters with non-

humans while in the wilder places of our homelands (including some not so wild). In each case, these encounters were witnessed by multiple individuals. These testimonies were from highly reliable sources, including a Georgia Magistrate and a Federal law enforcement agent. In one case there were more than a dozen people with the same testimony. Mass hysteria? I think not... not with the very people gathered for research into the being who confronted the group.

We have heard from filmmakers, costume makers, anatomists, professors, doctors and creature creators. We have even heard from Academy Award winners, all saying the same thing… "This being is NOT a human in a suit…"

Now, we have added one more dimension to the scene… irrefutable… reproducible… measurable evidence. The Intermembral Index is not something dreamed up by a field researcher to answer his own question, but is a subject taught in and is a product of our universities! It is part of any course taught on anatomy and body structure.

Have not these criteria forever been dangle before our eyes as the penultimate in evidence? "Show us this kind of evidence and you will have us convinced," they said to us. Reproducible… measurable… irrefutable… There it is, is it not?

Have we not used it to show that some films were bogus as well as to verify others? Haven't we always known that some films were real and true while others were simply someone's idea of a joke, I suppose… whatever they were, they were NOT real evidence!

I think it would take someone with a real grudge to not be able to accept the testimony of Bill Munns, John Chambers, Rueben Steindorff, Dr. Andrew Nelson, Dr. Curt Nelson, Dr. Scott Lund and Dr. Jeff Meldrum. Then more so from Emmy award winning animator, Joe Russo, Peter Brooke from "Jim Henson's Creature Shop" … I believe those people have some expertise in creating make believe. Mr. Brooke simply stated, referring to the being in the PGF… "Such costumes *did not exist* in the 60's…"

It was Dr. Jeff Meldrum, Professor of Anatomy and Paleontology at Idaho State University in Pocatello, Idaho who stated, "Look at the evidence… there is something leaving large footprints all over our country for us to find!"

Was John Chambers, 1968 Academy Award winning costume designer for "Planet of the Apes" being untruthful when he stated, "… The shoulder blade is clearly visible and moves during the walk and look back… *I don't know how they could do that…*"

On and on the testimonies came… Richard Stewart Taylor… the incident at Mile Post 14 on the G-O Road… a very special night in Georgia. Most essentially, those who spoke of the casts they made of the posteriors of some large, hairy beings…

There was more, of course. Michael Beers testified of his witnessed account that occurred in the Blue Mountains of Washington. Kyle Gibson testified of his encounter leading to his being frozen in place for a time on a lonely lakeshore in north-central Washington. Even more, there was Sgt. Todd Neiss with his witnessed account while on active duty with the US Army in coastal Oregon… and who can forget Click and his interaction with a pair of testifiers?

Finally, there is the evidence supplied by Barb Shupe and her group. She very carefully records the findings and leaves their interpretation to others. She is careful to make faithful investigations and representations of each event and her word is her bond.

This is the evidence as presented by those who purport the existence of a large, hairy being in our wild places. The only things presented in opposition are innuendo, assumption and conjecture. "My brother-in-law said…" has never been admissible evidence, but here, it is the backbone of their case.

NEVER accept unsubstantiated claims as any kind of proof… it is not even valid testimony, let alone proof. Of course, when we denigrate their "testimony", they immediately shift and attack us!

Remember, we have the preponderance of evidence on our side of the issue as well as the highest quality evidence. We have the testimony of quality people in great numbers and corroborated by other witnesses.

In conclusion on this summation, let me offer an anecdote. When the Covid vaccine first became available, it was a question as to whether it was a proper thing to do to have it administered. I, like many others, was undecided, so I did my homework, got with professionals I trusted and decided what was better for me and followed that route. When I mentioned what I had done, a person referred to me as a "sheeple", a coined word then being used to belittle those who did not agree with those who were militantly opposed to receiving the vaccine.

I stopped him and replied, "No, my friend, I did my research and decided what was best for ME and acted accordingly... that does not make one a sheeple... what makes one a sheeple is automatically rejecting something without research and parroting a made-up word coined by someone else like a sheep..." End of conversation... The same thing applies here... make sure that those with a different opinion are basing that opinion on fact and research, not just being a sheeple by parroting another's words!

Don't Box Yourself In!!

Some Logical Thinking Concerning the P-G Film

1. Why Travel Over 600 Miles from Home to Film a Hoax?

We have the Cascade Mountains in our backyard and the Blue Mountains in our front yard... both are literal hot spots for sightings.

Murphy's Law of Remote Location: The further you are from home, the higher the odds you've forgotten something essential!

Corollary to Murphy's Law: The further you are from home, the higher the odds that the item forgotten will not be available locally!!

If filmed at the end of some non-descript logging road, who would ever know?

2. Why Spend 3 Weeks Waiting to Film?

- Spend all Day Riding Horseback

- Spend all Night Driving Roads

- Camping in Undeveloped Campground

- Tending Stock and Keeping Life Running

3. Why Use Tag End of Film Roll?

- 76′ of 100′ Roll of Film Had Been Exposed Filming Bob and His Packhorse with the Flora of the Region

- Only 24′ of Film Left for Hoax Shoot

- Makes ID of the Site Irrefutable, But Does Nothing to Create a Film of Worth

4. Why Would You Choose a Site 3½ Miles from Nearest Road?

- Why Choose a Spot Accessible ONLY by Foot or by Horseback?

- Remember Murphy's Law?

    - It Compounds Exponentially When You Are Remote from a Remote Spot

5. Why Would You Film Next to a Noisy Creek?

- How Would You Communicate with Your Actor?

- Banks Are Tricky

- All Communication Must be by Hand Signal

- Necessary to Navigate Across Rough Ground and Water

6. Why Film the Entire Scene from Behind?

- Zero Communication with Actor

- Zero Control of Route and Speed

- Zero Ability to Adjust Anything in Process

- Photographer is Constantly Chasing Actor to Get the Scene

  - In a Costume Suit the Back is the WORST Part to Make Look Right – "NO ONE Films a Suit from the Rear if he has a Choice in the Matter" – Bill Munns

7. Why Shoot Only ONE Take??

- He KNEW he Ran Out of Film.

- He Knew He Fell at the Creek and does not Know if Camera Worked Subsequently

- He Had Driven 600 Miles from Home to Shoot This Hoax with Three Men and Three Horses

Does any of this make sense to ANYONE?

Any ONE of These Seven is Enough to Create Doubt of a Hoax.

- Putting all Seven together is the nail in the hoax idea's heart!!!

How Does the Nay-Sayer Respond to These Points?

He Doesn't… He ignores them…

And if you press the point??

He Attacks YOU!!

Bob Heronimous' Published Words:

1. I drove down to Orleans and we drove about four or five miles up the Bluff Creek Road. I put on the suit, walked across the gravel while Roger filmed it. I then took off the suit and drove home.

A. We took only one shot.

B. The suit was made from the hide of a red horse (later changed to black)

It was a two-piece suit… Bottom was on fishing waders and the top was pulled on over my head.

D. I wore football pads to beef it up.

E. Roger owed me money and everyone else made money off this, so I should too!

- *Show me the hem of the top pulled over the fishing waders!!*

- *Show me the football pads…*

Do you think he drove 14 hours to reach them, and then turned back around and drove 14 hours

home?

We Rest - Or Do We?

Yes, we have presented a strong case for our clients… but is it enough? In fact, I think not. To this point we have presented an overwhelming case for the fact that a non-human bi-pedal creature exists, but I think we can go one step further and state that, in view of the new, scientifically based Intermembral Index evidence, there exists, in our wild places, a creature known by different names in different places. He is Swamp Ape in Florida, Booger in much of the Southeast region and Bigfoot across much of the land, but he is Sasquatch in the West! He exists and his Intermembral Index proves that absolutely!

If this were not the case, we would not have this plethora of evidence, all with the same IMI figures! To suggest there are people all across the nation emulating the hoaxer, Dyer, and trying to sensationalize the situation and all are choosing the identical IMI in doing so, absolutely boggles the mind! We have so many identical cases for one simple reason… and it's the same reason humans have an IMI of 72… it is just what it IS!

Ladies and Gentlemen… Researchers, Investigators, Experiencers, Believers, Knowers, Wonderers, Doubters, Scoffers, Haters and Lovers, please allow me to take this moment and introduce you to…

Sasquatch… *Homo sapiens cognati*